Matter, Antimatter and Dark Matter

Matter, Antimatter and Dark Matter

Proceedings of the Second International Workshop
Trento, Italy   29 — 30 October 2001

editors

Roberto Battiston
Bruna Bertucci

I.N.F.N and Università di Perugia, Italy

World Scientific
New Jersey • London • Singapore • Hong Kong

Published by

World Scientific Publishing Co. Pte. Ltd.
P O Box 128, Farrer Road, Singapore 912805
USA office: Suite 1B, 1060 Main Street, River Edge, NJ 07661
UK office: 57 Shelton Street, Covent Garden, London WC2H 9HE

British Library Cataloguing-in-Publication Data
A catalogue record for this book is available from the British Library.

MATTER, ANTI-MATTER AND DARK MATTER
Proceedings of the Second International Workshop

Copyright © 2002 by World Scientific Publishing Co. Pte. Ltd.

All rights reserved. This book, or parts thereof, may not be reproduced in any form or by any means, electronic or mechanical, including photocopying, recording or any information storage and retrieval system now known or to be invented, without written permission from the Publisher.

For photocopying of material in this volume, please pay a copying fee through the Copyright Clearance Center, Inc., 222 Rosewood Drive, Danvers, MA 01923, USA. In this case permission to photocopy is not required from the publisher.

ISBN 981-238-118-X

Printed by FuIsland Offset Printing (S) Pte Ltd, Singapore

II International Workshop on Matter, Anti-Matter and Dark Matter

Trento, Italy, 28-29 October 2001

Organizing Committee

Roberto Battiston (University and INFN of Perugia)
Giuseppe Battistoni (INFN of Milano)
Franco Cervelli (INFN of Pisa)
Andrea Contin (University and INFN of Bologna)
Marina Gibilisco (ECT*)
Renzo Leonardi (ECT* and University of Trento)

Workshop sponsors

INFN
ASI
ECT*
Dipartimento di Fisica di Perugia

Editor's Foreword

The 2nd International Workshop on Matter, anti-Matter and dark Matter was held at Villa Tambosi (Trento), the ECT* headquarters, on October 28-29, 2001. This meeting is the second of a series, the first workshop was held in 1998, aimed to discuss open issues and trends in astro-particle physics in a inspirational and relaxed enviroment, as the one offered by ECT*.

A major goal of these meetings is to generate a congenial atmosphere in which partecipants feel comfortable discussing new ideas in addition to the results already achieved: the different subjects are reviewed in invited talks and a close interaction between all partecipants is guaranteed by the small size of the meeting.

The present Workshop was devoted to the progress in Cosmic Rays physics following the recent results obtained by balloon, satellite and underground experiments.

The following topics have been reviewed in four plenary sessions: composition and propagation of cosmic rays, trapping of charged particles in the earth's magnetic field, atmospheric neutrinos and high energy photons measurements in space. A round table with representatives of the Italian and European Space Agencies (G.Bignami, S.Vitale), and Italian and Russian research centers (R.Battiston, L.Scarsi, M.Khrenov, M. Panasyuk) closed the meeting illustrating the perspectives of astroparticle physics in space as seen in different scientific communities.

A total of 66 physicists, from 10 countries, attended the meeting: we particularly thank the speakers who contributed with competence and enthusiasm to the scientific success of the workshop.

Special thanks go to all the people involved in the organization of the Workshop, in particular to Marina Gibilisco, local member of Organizing Committee, and Cristina Costa from the ECT* secretariat. The technical competence of Gianluca Scolieri (University of Perugia) allowed a nearly real-time publishing of the transparencies of all the talks on the WEB.

Perugia, March 2002

Roberto Battiston
Bruna Bertucci

List of Participants

Alpat, B.	Behcet.Alpat@pg.infn.it
Ambrosi, G.	Giovanni.Ambrosi@pg.infn.it
Azzarello, P.	Philipp.Azzarello@cern.ch
Battiston, R.	battisto@krenet.it
Battistoni, G.	giuseppe.battistoni@mi.infn.it
Bertucci, B.	Bruna.Bertucci@pg.infn.it
Bignami ,G.	bignami@asi.it
Biland, A.	biland@particle.phys.ethz.ch
Borgia, B.	bruno.borgia@roma1.infn.it
Bosio, C.	carlo.bosio@roma1.infn.it
Buenerd, M.	buenerd@in2p3.fr
Casadei, D.	Diego.Casadei@bo.infn.it
Casaus, J.	Jorge.Casaus@ciemat.es
Casolino, M.	casolino@roma2.infn.it
Cecchi, C.	claudia.cecchi@pg.infn.it
Cei, F.	fabrizio.cei@pi.infn.it
Cervelli, F.	cervelli@pi.infn.it
Chang, Y.H.	Yuan-Hann.Chang@cern.ch
Chen, G.	Gang.Chen@cern.ch
Choumilov, E.	Evgueni.Choumilov@cern.ch
Choutko, V.	v.choutko@cern.ch
Contin, A.	contin@bo.infn.it
Cristinziani,M	Markus.Cristinziani@cern.ch
Derome, L.	laurent.derome@isn.in2p3.fr
Durante, M.	durante@na.infn.it
Echenard, B.	bertrand.echenard@cern.ch
Esposito, G.	gennaro.esposito@pg.infn.it
Fiandrini, E.	emanuele.fiandrini@pg.infn.it
Gentile, S.	simonetta.gentile@roma1.infn.it
Gibilisco, M.	marina@ect.it
Girard, L.	loic.girard@ lapp.in2p3.fr
Heynderickx, D.	D.Heynderickx@bira-iasb.oma.be
Iannucci, A.	iannucci@roma2.infn.it
Ionica, R.	romeo.ionica@pg.infn.it
Khrenov, B.	khrenov@eas.npi.msu.ru
Kirn, T.	Thomas.Kirn@physik.rwth-aachen.de
Klimentov, A.	alexei.klimentov@cern.ch
Lamanna, G.	Giovanni.Lamanna@cern.ch

Leluc, C.	catherine.leluc@physics.unige.ch
Lubrano, P.	pasquale.lubrano@pg.infn.it
Maestro, P.	maestro@pi.infn.it
Malinine, A.	Alexandre.Malinine@cern.ch
Marrocchesi, P.S.	marrocchesi@pi.infn.it
Masiero, A.	masiero@sissa.it
Mihul, A.	Alexandru.Mihul@cern.ch
Mikhailov, V.	vladimir@space.mephi.ru
Morselli, A.	aldo.morselli@roma2.infn.it
Panasyuk, M.	panasyuk@sinp.msu.ru
Piron, F.	piron@in2p3.fr
Pohl, M.	Martin.Pohl@cern.ch
Plyaskin, V.	vassili.plyaskine@cern.ch
Produit, N.	produit@mbx.unige.ch
Sanuki, T.	sanuki@phys.s.u-tokyo.ac.jp
Sbarra, C.	sbarra@bo.infn.it
Shiozawa, M.	masato@icrkm4.icrr.u-tokyo.ac.jp
Siedenburg, T.	thorsten.siedenburg@cern.ch
Sparvoli, R.	roberta.sparvoli@roma2.infn.it
Stanev, T.	stanev@bartol.udel.edu
Suter, H.	Henry.Suter@cern.ch
Tavani, M.	tavani@ifctr.mi.cnr.it
Ulbricht, J.	ulbricht@particle.phys.ethz.ch
Ullio, P.	ullio@sissa.it
Valle, G.	giada.valle@pi.infn.it
Vialle, J.P.	vialle@lapp.in2p3.fr
Vietri, M.	vietri@fis.uniroma3.it
Vissani, F.	vissani@lngs.infn.it
Zuccon, P.	paolo.zuccon@pg.infn.it

CONTENTS

Committees and Sponsors — v

Editor's Foreword — vii

List of Participants — ix

Recent Measurements on Cosmic Rays Spectra and Composition
Chairperson: M. I. Panasyuk

The Alpha Magnetic Spectrometer, A Particle Physics Experiment in Space — 1
   R. Battiston

Review on Precision Measurements of High Energy Hadrons — 15
   J. Casaus

Review of Precision Measurements of High Energy Electrons — 25
   B. Bertucci

A Monte Carlo Simulation of the Cosmic Rays Interactions with the Earth's Atmosphere — 37
   P. Zuccon

Review of Balloons Muon Measurement in the Atmosphere — 47
   T. Sanuki

An Analytical Solution of the Cosmic Rays Transport Equation in the Presence of the Geomagnetic Field — 57
   M. Gibilisco

Interaction of Cosmic Rays with the Geomagnetic Field
Chairperson: G. Battistoni

Leptons with E > 200 MeV Trapped in the South Atlantic Anomaly — 67
   E. Fiandrini and G. Esposito

Simulation of Atmospheric Secondary Hadron and Lepton Flux from Satellite to Underground Experiments — 77
   M. Buénerd

Review on Modelling of the Radiation Belts — 87
   D. Heynderickx

Low Energy Solar and Galactic Cosmic Rays at 1 AU — 97
   M. Casolino

Low Energy Electron and Positron Spectra in the Earth Orbit Measured by MARIA-2 Instrument — 107
   V. V. Mikhailov

The Trapped Anomalous Component of the Cosmic Rays: The Short Overview of Experiments M. I. Panasyuk	117
Biological Effects of Cosmic Radiation in Low-Earth Orbit M. Durante	125
Neutrinos as Dark Matter Candidates A. Masiero and S. Pascoli	135

Recent Developments on Atmospheric Neutrinos
Chairperson: F. Cervelli

Simulation of Particle Fluxes in the Earth's Vicinity V. Plyaskin	145
Calculation of Secondary Particles in Atmosphere and Hadronic Interactions G. Battistoni, A. Ferrari and P. R. Sala	155
Massive Neutrinos and Theoretical Developments A. Strumia and F. Vissani	167
Neutrinos from Supernovae: Experimental Status F. Cei	177

Dark Matter and Gamma Rays
Chairperson: B. Bertucci

Searches for Dark Matter Particles Through Cosmic Ray Measurements P. Ullio	189
Integral: A Gamma-Ray Observatory N. Produit	199
The AGILE Mission and Gamma-Ray Astrophysics M. Tavani	211
Pulsars, Blazars and Dark Matter with AMS M. Pohl	221
Cosmic Photon and Positron Spectra Measurements Modelling with the AMS-02 Detector at ISS V. Choutko, G. Lamanna and A. Malinin	229
Dark Matter Search with Gamma Rays: The Experiments EGRET and GLAST A. Morselli	241

Recent Meaurements on Cosmic Rays Spectra and Composition

Chairperson: *M. I. Panasyuk*

THE ALPHA MAGNETIC SPECTROMETER, A PARTICLE PHYSICS EXPERIMENT IN SPACE

ROBERTO BATTISTON *

Dipartimento di Fisica and Sezione INFN
Via Pascoli, Perugia Italy

The Alpha Magnetic Spectrometer (AMS) is a state of the art detector for the extraterrestrial study of matter, antimatter and missing matter. During the STS-91 precursor flight in may 1998 AMS collected nearly 100 millions of Cosmic Rays on Low Earth Orbit, measuring with high accuracy their composition. We review the results on the flux of proton, electron, positron and helium. Analysis of the under cutoff spectra indicates the existence of a new type of belts of energetic trapped particles characterized by a dominance of positrons versus electrons. AMS is currently being refurbished for a three year mission on the International Space Station where the its sensitivity to rare events will be increased by three to four orders of magnitude.

Keywords: antimatter; dark matter; space station; magnetic spectrometer; belts; cosmic rays

1. Introduction

The disappearence of the antimatter [1,2,3] and the presence at all scales in our universe of a non luminous components of matter (dark matter)[4,5] are two of the most intriguing misteries in our current understanding of the structure of the Universe.

To study these problems, a high energy physics experiment, the Alpha Magnetic Spectrometer (AMS)[6], is scheduled for installation on the International Space Station in 2004. Goal of AMS is to perform a three year long measurement, with the highest accuracy, of the composition of Cosmic Rays in the rigidity range 0,1 GV to several TV. In preparation for this long duration mission AMS flew a ten days precursor mission on board of the space shuttle Discovery mission STS-91 in June 1998. This high statistics measurement of CR in space, enabled, for the first time, the systematic study the behaviour of primary CR near Earth in the rigidity interval from $0,1~GV$ to $200~GV$, at all longitudes and latitudes up to $\pm51.7^o$. In this paper we present some relevant results obtained by AMS during the precursor mission. We also report the observation of high energy radiation belts in the near Earth region and on their composition, which shows remarkable differences with previously observed belts of trapped particles around our planet.

---

*Email: battisto@krenet.it

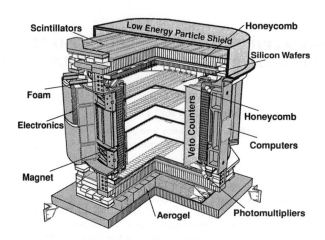

Fig. 1. AMS on the Discovery STS 91 precursor flight, June 1998.

2. The AMS experiment on the STS-91 mission

Search of antimatter requires the capability to identify with the highest degree of confidence, the mass of particle traversing the experiment together with the absolute value and the sign of its electric charge.

The AMS configuration flown in 1998 on the Shuttle Discovery (Fig.1) includes a permanent Magnet, Anticounter (ACC) and Time of Flight (ToF) scintillator systems, a large area, high accuracy Silicon Tracker and an Areogel Threshold Cherenkov counter. The magnet is based on recent advancements in permanent magnetic material and technology which make it possible to use very high grade Nd-Fe-B to build a permanent magnet with $BL^2 = 0.15\ Tm^2$ weighting ≤ 2 tons. A charged particle traversing the spectrometer triggers the experiment through the ToF system, which measures the particle velocity with a resolution of $\sim 120\ ps$ over a distance of $\sim 1.4\ m$[11].

The pattern recognition and tracking is performed using the large area ($\sim 7\ m^2$), high accuracy Silicon Tracker[7,9], which, for the Space Station mission, will be covered with 2300, high purity, double sided, 300 μm thick silicon wafers[12], following the technology developed in Italy by INFN for the Aleph[10] and L3[8] vertex detectors at LEP. The active area of the AMS Silicon Tracker is about an order of magnitude larger than in the case of the microstrip silicon detectors presently installed at high energy Colliders. AMS is the first high energy spectrometer based only a high precision multilayer Silicon Tracker.

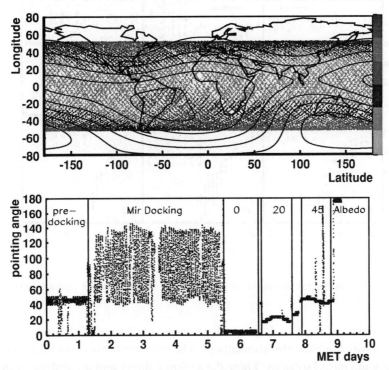

Fig. 2. (a)Shuttle orbits during the STS-91 mission: the different intensities correspond to different C.R. rates, lower at the equator and larger close to the poles and (b) shuttle attitudes during the mission as a function of the Mission Elapsed Time (MET).

The momentum resolution for AMS on on the precursor mission was about ($\frac{\Delta p}{p} \sim 7\%$) at 10 GV, reaching ($\frac{\Delta p}{p} \sim 100\%$) at about 500 GV.

Four ToF scintillators layers and up to eight Silicon Tracker layers measure $\frac{dE}{dx}$, allowing a multiple determination of the absolute value of the particle charge.

By combining the various measurement it is then possible to determine the type of particle traversing the magnet and identify interesting particles with a background rejection which for anti-matter searches is expected to reach one part in 10 billions.

During the period june 2nd to june 12th, 1998 the Shuttle Discovery has performed 154 orbits at an inclination $51.7°$ and at an altitude varying between 390 to 350 km. During the mission AMS collected a total of about 100 Million triggers, at various Shuttle attitudes (Fig.2). In the Figure one notice the period of Shuttle to Mir docking when the Shuttle attitudes are rapidly changing with time.

Almost all results published so far[13,14,15] were obtained with data collected during well defined attitude periods with AMS pointing at $0°, 20°$ and $45°$ with respect to zenith (deep space).

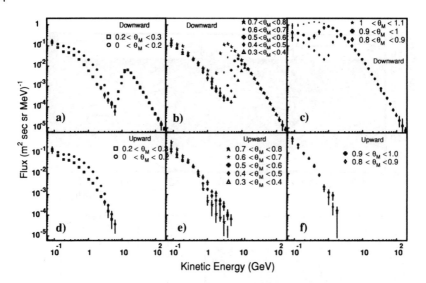

Fig. 3. Proton spectra measured by AMS for different geomagnetic latitude intervals.

These data are the first high quality CR data collected with a magnetic spectrometer located outside the atmosphere. The measurements cover all geomagnetic longitudes and most latitudes. These data allow a direct and accurate measurement of the CR composition and spectra, as well as a systematic study of the effects of the geomagnetic field.

The measurement of the proton flux as a function of the geomagnetic latitude (Fig. 3a-3c), shows that, in addition to the primary CR spectrum visible above the geomagnetic cutoff, there is a substantial second spectrum, extending to much lower energy and exhibiting some significant latitude dependence close to the equator. These particles cannot come from the deep space, they are on forbidden orbits, but are produced in the interaction of the primary CR with the top layers of the atmosphere. A characteristic of the second spectrum is that it is up-down symmetric (Fig. 3d-3f).

Second spectra with similar geomagnetic latitude dependence have been detected by AMS in the low energy region of the spectra of e^-, e^+[16], D [17] and, although with weaker intensity, ^3He. Fig. 4 and 5 show the results for electrons, positrons and Helium. These results on the second spectra are discussed later in the paper.

Adding all data collected above the geomagnetic cutoff it is possible to obtain a precise estimate of the primary CR differential flux. Parametrizing the omnidirectional CR flux as $\Phi(R) = \Phi_o R^\gamma$ (R in GV) we obtain the results reported in Table 1.

It it interesting to compare the AMS measurement of the primary fluxes with

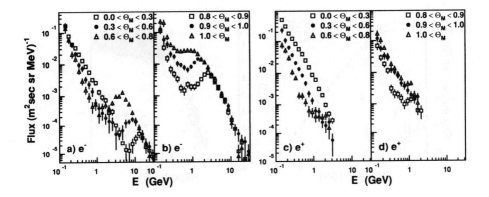

Fig. 4. Electrons and positron spectra measured by AMS during the STS91 flight.

Table 1. AMS results on the parametrization of proton and helium primary flux.

Proton flux	
γ	2.78 ± 0.009 (fit) ± 0.019 (syst)
Φ_o	17.1 ± 0.15(fit) ± 1.3(syst) $\pm 1.5(\gamma) GV^{2.78}(m^2 s\, sr MeV)^{-1}$
Helium flux	
γ	2.740 ± 0.010(fit)± 0.016(syst)
Φ_o	2.52 ± 0.09(fit)± 0.13 (syst) $\pm 0.14(\gamma) GV^{2.78}(m^2 s\, sr MeV)^{-1}$

previous results obtained with stratospheric balloons[18,19,20,21,22]. Fig.7 shows the comparison for the proton spectrum, multiplied by $E^{2.5}$. The improved statistical significance and the wider energy interval covered by AMS data is evident: thanks to the improved accuracy obtained with only few days in space, it is possible to clarify the situation resulting from the data published over the last 15 years by the various Collaborations using different implementations of the NASA New-Mexico spectrometer[19,20,21,22] and by the Bess Collaboration [18].

Similar consideration apply for the comparison of the measurement of Helium primary flux (Fig.7).

Both for protons as well as for Helium, AMS show a nice agreement with the measurement of the Bess Collaboration[18], although our data have a smaller statistical error and extends over a wider energy interval. Using the large He sample collected by AMS a search for anti-He candidates has also been performed. Within 2.3 Millions He events no anti-He candidates have been found, up to a rigidity of 140 GV.

Assuming identical He and anti-He spectra we obtain a model independent upper

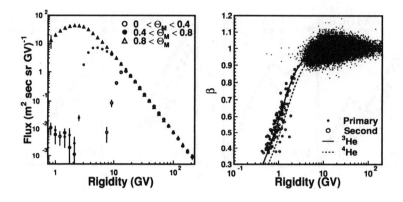

Fig. 5. Helium spectra measured by AMS during the STS91 flight.

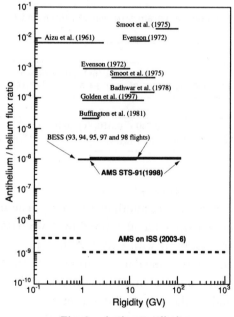

Fig. 6. Antimatter limits.

limit of 1.110^{-6} over the rigidity interval 1 to 140 GV, which can be compared to previous results (Fig.6).

3. Observation of high energy particle belts

The trapping of charged particles in the quasi dipolar earth magnetic field is a classical problem, which has been studied in great detail[27] following Van Allen

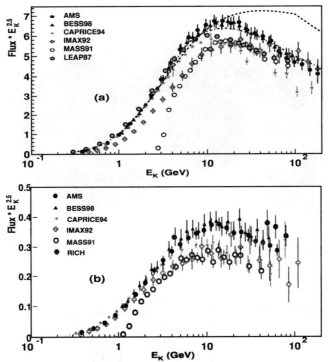

Fig. 7. (a) Primary proton flux measured by AMS and compared with existing balloons measurements. The lines are parametrizations of the primary cosmic rays used in atmospheric ν flux calculation: dashed line HPPK [23], dot-dashed line Bartol group [24]; (b) primary He flux measured by AMS and compared with existing balloons measurements.

observations in 1958 [29]. The basic physical mechanism is well understood. For sufficiently low rigidities, the trapped particles spiralize along orbits defining shells surrounding our planet.

These shells are shaped along the magnetic field lines and are roughly symmetric in latitude with respect to the geomagnetic equator (Fig.8). The motion of a trapped particle can be separated in three components, the revolution around the guiding center or gyration, the bouncing between mirror points located ≈ symmetrically with respect to the geomagnetic equator (magnetic bottle), and a longitudinal drift around the earth. The geometrical locations defined by the orbits of trapped particles are called shells. A shell can be univocally determined by two parameters. For example a pair of variables are L, the distance of the shell at the equator measured in unit of the Earth radius ($R_\oplus$), and B_{mir}, the value of the magnetic field at the point where the particles reverse their motion (mirror point)[30]. Depending on the shell, B_{mir} can be locally very deep the atmosphere (it can be below the earth crust). Shells which are characterized by these value of B_{mir} cannot trap the particles, since they are lost within one or few bounces across the magnetic equator.

A particle belonging to a shell will remain on the same shell until it is disturbed

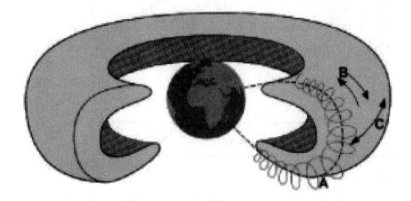

Fig. 8. Motion of charged particles in the geomagnetic belts. A) gyration B) bouncing C) drift.

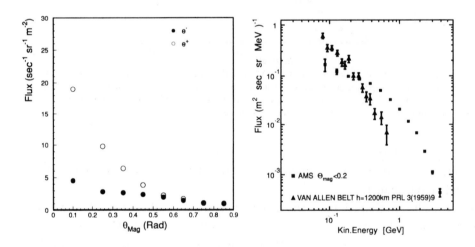

Fig. 9. (a) $\frac{e^+}{e^-}$ ratio inside the belts observed by AMS, as a function of the geomagnetic latitude; (b) comparison among a typical Van Allen belt proton spectrum and equatorial AMS belts proton spectrum.

by (a) interaction with the top layers of the atmosphere or other particles or (b) interaction with electrical or magnetic variable field.

Conversely, primary cosmic rays coming from deep space cannot enter a shell unless their trajectories are disturbed by some interaction with matter or fields. The existence of the shells is the result of the equilibrium between two mechanisms: some contributing to fill the shells with new particles and others removing some of the trapped particles.

Table 2. Different types of particle belts around the Earth.

Belt type	Particle type	Rigidity [MeV/n]	Filling mechanisms	L	Residence time [d]
Van Allen (inner)	p e^-	$0.1 - 100$ $0.01 - 1$	$n \to pe^-\bar{\nu}_e$, external belts	< 2.5	$10 - 1000$
Van Allen (outer)	e^- p	$1 - 10$ $0.1 - 1$	solar wind	> 2.5	$1 - 10$
SAMPEX	$N^{+x}, O^{+x},$ Ne^{+x}	10 $10 - 100$	Anomalous CR	2	$10 - 100$
AMS	p e^- e^+ ^3He	$100 - 1000$ $100 - 1000$ $100 - 1000$ $100 - 1000$	primary CR interacting with the atmosphere	≤ 1.15	$10^{-6} - 10^{-4}$

If the dynamics of the particles trapped is well understood, the mechanisms contributing to shell stability are much less understood. They involve: interaction of high energy CR with the atmosphere creating neutrons which decays in flight, $n \longrightarrow p + e^- + \bar{\nu}_e + 782 KeV$, filling the belts (CRAND mechanism[28]), instabilites due to solar storms, as well as other types of magnetic and electric instabilities. It should be pointed out, however, that the mechanisms proposed are compatible with the observed dominance of protons and electrons in the Van Allen belts.

The shell can be classified by their composition and location. The original Van Allen belts contain only proton and electrons and extend to very large distance from the earth, up to $L \approx 6$. Van Allen belts are divided into inner and outer belts, since there is a dip in the particle flux intensitiy at about 2.5 L. During the last 20 years, there have been reports of the observation of a low flux of trapped ions, mainly He and O, with traces of C e N, and having energies of a few MeV/n and $L = 3 - 4$. These particles are extracted from the upper layers of the atmosphere during solar storms. More recently, nearly 40 years after Van Allen discovery, the analysis of SAMPEX data[32] has shown the existence of belts included in the inner Van Allen belts, containing heavier nuclei like N, O, Ne with rigidities of the order of $10/MeV$.

The SAMPEX belts are different from the Van Allen belts mainly because of their composition due to a different the filling mechanism, which is likely due to the interaction of the so called Anomalous Cosmic Rays with the Earth atmosphere[33,31].

The belts observed by AMS are different in composition since they also contain a large fraction of positrons, but also deuterium and ^3He. These particles have not been observed in the Van Allen or SAMPEX belts. Particularly striking is the abundance of positrons versus electrons (Fig.4), with a ratio exceeding a factor of four in the equatorial region (Fig. 9a).

AMS observed shells with $L \leq 1.15$, well below the inner Van Allen belts. In the

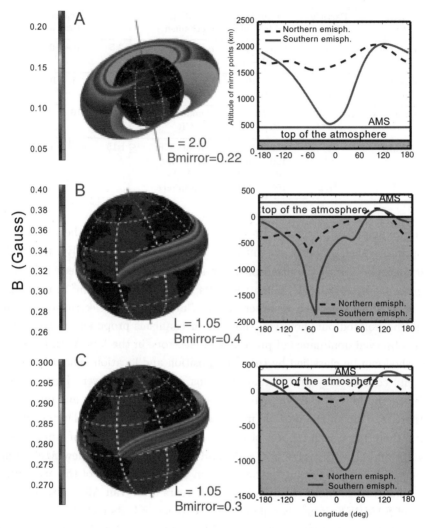

Fig. 10. Van Allen versus AMS belts. (A) Van Allen belts have high L values, B_{mirr} is located mainly above AMS orbits, particles weakly interact with the atmosphere and have lifetimes ranging from days to months. AMS belts have L $\leq$ 1.5, and depending on the value of the B_{mirr} their lifetime ranges from fractions of a second (B) to several seconds (C).

belts studied by AMS the observed proton spectrum is harder (Fig. 9b) than in the case of Van Allen belts. This can be understood since their location is closer to the earth and the particles do experience a stronger trapping field. Another difference with the Van Allen belts is the residence time of the trapped particles, computed using computer based tracing techniques, which is in the region of seconds and not days or weeks. These shells cannot be observed by stratospheric balloons, since

Table 3. Physics capabilities of AMS-02 after three years on the ISS

Elements	Sensitivity	(Now)	Rigidity Range(GV)	Physics
e^+	10^7	$(\sim 10^3)$	$0.1 - 200$	↑
$\bar{p}$	10^6	$(\sim 2\ 10^3)$	$0.5 - 100$	Dark Matter
γ			$1 - 300$	↓
$\bar{H}e/He$	$\frac{1}{10^9}$	$(\frac{1}{10^6})$	$1 - 200$	Antimatter
$\bar{C}/C$	$\frac{1}{10^8}$	$(\frac{1}{10^4})$	$1 - 200$	CP, GUT, EW
D	10^9		$1 - 20$	↑
$^3He/\,^4He$	10^9		$1 - 10$	Astrophysics
$^{10}Be/\,^9Be$	2%		$1 - 10$	↓

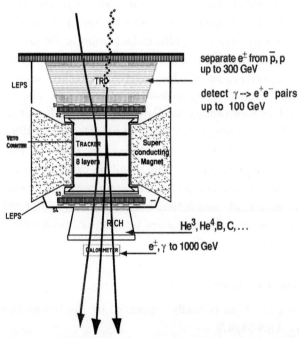

Fig. 11. Configuration of AMS on the ISS for the three years mission scheduled on UF4 in 2004.

their mirror fields are above the atmosphere except in correspondence of the South Atlantic Anomaly. It follows that the observed particles do not belong to the various types of albedo particles reported in the past by experiments on balloons.

In Table 2 we summarize the main features of the different type of belts identified during the last 40 years. As we can see the situation is very varied, corresponding to different filling mechanisms. Since we are dealing with continuous distributions, the reported intervals (rigidity, L, residence time) should be taken as typical order of magnitudes. In Fig.10, we compare the structure of the AMS belts to the Van Allen belts as well as the dependence of the mirror field altitude with the longitude.

4. Conclusions

The first mission of the Alpha Magnetic Spectrometer, although lasting only ten days, has been scientifically very rewarding, allowing for the first time a very detailed measurement of high energy cosmic rays outside the atmosphere. In addition to the most accurate measurements obtained so far for the primary flux of $p, e^+, e^-, D, ^3He$ and ^4He spectra over most of the earth surface, these results have shown the existence of a substantial second spectrum of high energy particles trapped within low altitude belts. These new belts have a very characteristic composition, dominated by positively charged particles, mainly p, e^+ and D. Their existence should be taken into account when calculating radiation doses for astronauts on the ISS or background rates for low orbit satellites.

AMS is currently being refurbished to be ready for a three years mission with UF4 in 2004. A stronger magnetic field from a superconducting magnet, $B = 0, 7\,T$, a fully equipped Silicon Tracker, together with three powerful particle identification detectors, a Transition Radiation Detector, a Ring Imaging Cherenkov (RICH) detector and an Electromagnetic Calorimeter, will allow precise particle identification up to $O(TeV)$ of energy (Fig.11). The physics capabilities of AMS after three years of exposure on the ISS are summarized in Table 3. AMS will be the only large acceptance magnetic facility which will be exposed for long time in space. It will allow a measurements of the flux and composition of Cosmic Rays with an accuracy orders of magnitude better than before. The large improvement in sensitivity given by this new instrument, will allow us to enter into a totally new domain to explore the unknown.

5. Acknowledgment

This work has been partially supported by the Italian Space Agency (ASI) under contract ARS-98/47.

References

1. Steigmann, G., Ann. Rev. Astron. Astroph., **14** p. 339, 1976.
2. Kolb, E.W., Turner, M.S., Ann. Rev. Nucl. Part. Sci. **33** p. 645, 1983.
3. Peebles, P.J.E., Principles of Physical Cosmology, Princeton University Press, Princeton N.J., 1993.
4. Ellis, J. et al., Phys. Lett. **B214**, p. 403, 1988.
5. Turner, M.S., Wilzek, F., Phys. Rev. **D42**, p. 1001, 1990.
6. Ahlen, S. et al., Nucl. Inst. Meth. **A350**, p. 351, 1994.
7. Battiston, R., Nucl. Instr. Meth. (Proc. Suppl.) **B44**, p. 274, 1995.
8. Acciarri, M. et al., Nucl. Inst. Meth. **A289** p. 351-355, 1990.
9. Alcaraz, J. et al., Il Nuovo Cimento **112A**, p. 1325-1344, 1999.
10. Batignani, G. et al., Nucl. Inst. Meth. **A277** p. 147, 1989.
11. Alvisi, D. et al., Nucl. Inst. Meth. **A437** p. 212, 1999.

12. Produced at CSEM, SA Rue J. Duot 1, P.O. Box, CH-2007 Neuchatel, http://www.csem.ch.
13. AMS Collaboration, Alcaraz, J. et al., Phys. Lett. **B461**, p. 287, 2000.
14. AMS Collaboration, Alcaraz, J. et al., Phys. Lett. **B472**, p. 215, 2000.
15. AMS Collaboration, Alcaraz, J. et al., Phys. Lett. **B484**, p. 10, 2000.
16. AMS Collaboration, Alcaraz, J. et al., Phys. Lett. **B490**, p. 27, 2000.
17. Lamanna, G. PhD Thesis, University of Perugia, October 2000, unpublished.
18. BESS98, Sanuki, T. et al., astro-ph/0002481, 2000.
19. CAPRICE94, Boezio, M. et al., ApJ **518**, p. 457, 1999.
20. IMAX92, Menn, W. et al., The Astrophys. J. **533**, p. 281, 2000.
21. MASS91, Bellotti, R. et al., Phys. Rev. **D60**, p. 052002, 1999.
22. LEAP87, Seo, E. S. et al., ApJ **378**, p. 763, 1991.
23. HKKM, Honda, M. et al., Phys. Rev. **52**, p. 4985, 1995.
24. BARTOL, Lipari, P. and Stanev, T., Talk given at Now 2000 Conference, 2000.
25. Smoot, G. F. et al., Phys. Rev. Lett. **35**, p. 258-261, 1975; Steigman, G. et al., Ann. Rev. Astr. Ap. **14**, p. 399, 1976; Badhwar, G. et al., Nature **274**, p. 137, 1978; Buffington, A. et al., ApJ **248**, p. 1179–1193, 1981; Golden, R. L. et al., ApJ **479**, p. 992, 1997; Ormes, J. F. et al., ApJL **482**, p. 187, 1997; Saeki, T. et al., Phys. Lett. **B422**, p. 319, 1998.
26. Nozaki, M., OG.1.1.23, 26th ICRC, Salt Lake City, Utah, 1999.
27. For a recent review see Walt, M., Radiation Belts Models and Standards, AGU Geophysical Monograph 97, p. 1, 1997.
28. Singer, S. F., Phys. Rev. Lett. **1**, p. 181, 1958; Hess, W. N., Phys. Rev. Lett. **3**, p. 11, 1959; Kellogg, P., J. Geophys. Res. **65**, p. 2, 705, 1960; Vernov, S. N. et al., Soviet Physics, Doklady **4**, p. 154, 1959.
29. Van Allen, Ludwig, Ray, and McIlwain, IGY Satellite Series Number 3, **73**, Natl. Acad. Sci., Washington D.C., 1958; Van Allen, McIlwain, and Ludwig, J. Geophys. Research **64**, p. 271, 1959; Van Allen, J. A. and Frank L. A., Nature **183**, p. 430, 1959.
30. McIlwain, C. E., J. Geophys. Res. **66**, p. 3681–3691, 1961.
31. Mewaldt, R. A., Radiation Belts Models and Standards, AGU Geophysical Monograph **97**, p. 35, 1997.
32. Cook, W. R., IEEE Trans. Geosci. Remote Sensing **31**, p. 557–564, 1993.
33. Cummings, J. R. et al., Geophys. Res. Lett. **20**, p. 2003–2006, 1993; Cummings, J. R. et al., IEEE Trans. Nucl. Sci. **40**, p. 1459–1462, 1993.

REVIEW ON PRECISION MEASUREMENTS OF HIGH ENERGY HADRONS

J. CASAUS

Dpto. Fusión y Partículas Elementales, CIEMAT, Avda. Complutense 22
Madrid, E-28040, Spain
E-mail: Jorge.Casaus@ciemat.es

Precise measurements of high energy hadrons have been performed either on balloon-borne or space-borne experiments. The status of the present measurements on H and He, heavier nuclei, isotopes and antiprotons is separately reported. Implications of precise measurements within the framework of models for production and propagation of galactic cosmic rays is discussed. Near future experiments are expected to improve in a significant manner the collected statistics and the energy range covered by present experiments. The results thus obtained will validate current propagation models and accurately constrain their free parameters.

Keywords: Cosmic rays

1. Introduction

Hadrons are the main component in cosmic rays (CR). Measurements of the CR element abundances compared to those in the Solar System have helped to understand qualitatively the source composition and propagation properties of galactic CR. However, only precise measurements of the elemental and isotopic spectra of hadronic CR can validate specific models for their origin and propagation. Precise knowledge of the free parameters in these models is needed to predict the expected fluxes and, in turn, to reduce the uncertainties in the background for rare signal searches in CR which could have deep impact in fundamental physics. The current program in CR propagation calculations aims to describe all primary and secondary species in galactic CR (hadrons and leptons) and diffuse γ–ray background in a single model. The parameters in these models are the composition and injection spectra at CR sources and the galactic disc and halo properties which can be derived from measurements done in the heliosphere.

This review will concentrate on the results obtained by instruments endowed with good particle identification. Results from extensive air showers, where the identification of the primary hadron impinging upon the atmosphere is rather poor, are thus excluded. Among the measurements performed in the last 40 years, only most relevant results will be included.

2. Present Measurements

The present precise measurements of hadronic CR come from the results obtained by instruments mounted either on stratospheric balloons operating at a floating altitude of about 40 km or on space satellites. Their limited energy range and nuclear electric charge coverage arise from the short exposure and intrinsic instrumental limitations. We will consider separately H and He, heavier nuclei, isotope measurements and antiprotons.

2.1. *H and He*

Hydrogen and helium constitute about 99% of the hadronic CR. Precise knowledge of their fluxes is needed to calculate rare secondary CR yield such as antiprotons and positrons, to compute the diffuse gamma ray background spectrum and to define the expected fluxes of atmospheric neutrinos. Although extensively measured for decades, only recent precise measurements have been able to pin down the uncertainties in a limited energy range.

We can define three different energy ranges:

(1) $E \lesssim 100$ GeV/n :
Most precise measurements come from recent balloon-borne (BESS [1], IMAX [2], CAPRICE [3]) and space-borne (AMS [4,5]) magnetic spectrometers. The estimated uncertainties are about 5% for H and 10% for He.

(2) 100 GeV/n $\lesssim E \lesssim 1$ TeV/n :
The only existing measurements are those of Ryan et al. [6] using a balloon-borne calorimeter and, for He, the results obtained with a balloon-borne RICH counter [7]. The estimated uncertainty is about 25%.

(3) 1 TeV/n $\lesssim E \lesssim 1000$ TeV/n :
Only emulsion chamber experiments have been able to accumulate statistics in this range. The most extensive data sets come from JACEE [8] and RUNJOB [9]. The intrinsic difficulties for detector calibration together with the low statistics set the present uncertainties at the level of 25% – 50%.

Figure 1 shows a summary of the hydrogen spectrum measurements coming from a recent compilation by T.K. Gaisser [10] together with fits to data used to compute the cosmic ray induced atmospheric neutrino yield [11,12].

2.2. *Heavier nuclei*

Among light nuclei above helium, the CNO group are the most abundant and believed to be of primary origin, i.e., present at CR sources. Lithium, beryllium and boron are believed to be products of spallation of heavier nuclei in their path through the interstellar medium. The ratio B/C has been used to estimate the amount of matter traversed by CR since their acceleration. In specific models, this ratio defines the CR escape path length distribution (leaky-box model) or the

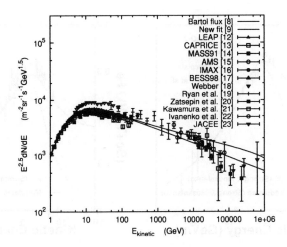

Fig. 1. Summary of recent measurements of the spectrum of CR hydrogen and fits to data used to compute atmospheric neutrino production. The reference numbering corresponds to a recent compilation by T.K. Gaisser.

spatial diffusion coefficient (diffusion models). Similarly, sub-iron elements (Sc–Ti–V), produced from primary iron spallation, can be used to determine the same quantity. Experimentally, the latter measurement is less precise due to statistical limitations of the sample and uncertainties in the charge assignment for heavy elements.

Most precise absolute fluxes of individual elements above helium have been obtained using space-borne instruments such as the C2 telescope onboard the HEAO-3 spacecraft [13] and the CRN detector on the Spacelab-2 mission onboard the Space Shuttle [14]. The joint energy range covered ranges from 0.6 GeV/n to 35 GeV/n for all species up to nickel (Z=28) and is extended up to 1 TeV/n for the more abundant elements. The uncertainties for energies $\lesssim$ 50 GeV/n are dominated by normalisation uncertainties of about 10%. For energies above this value, the measurement is strongly limited by statistics.

Boron to carbon ratio has been derived by HEAO-3 [13], Spacelab-2 [15] and, at lower energies, by a balloon-borne experiment [16] and by space-borne spectrometers such as the high energy isotope experiment on the ISEE 3 spacecraft [17]. Uncertainties in the ratio are $\sim$ 5% in the HEAO-3 energy range (0.6 – 35 GeV/n), increasing to $\sim$ 10% for the ISEE-3 measurements (0.1 – 0.2 GeV/n) and reaching $\sim$ 50% for the Spacelab-2 measurements (70 – 200 GeV/n).

Sub-iron to iron ratio has been obtained by HEAO-3 [13], by a balloon-borne experiment [16] and by space-borne spectrometers such as the CRIS experiment on the ACE spacecraft [18]. Uncertainties are $\sim$ 10% in the covered energy range (0.14 – 35 GeV/n).

Figure 2 shows these measurements together with the results obtained for a

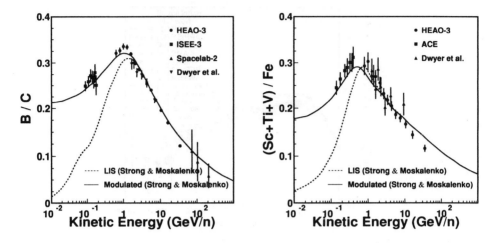

Fig. 2. Measurements of boron over carbon and sub-iron over iron ratios together with the expectations for a propagation model as described in the text. The dashed lines represent the ratio of local interstellar spectra, whereas the solid lines are the corresponding ratios after accounting for solar modulation effects. The references are described in the text.

propagation model including diffusion and reacceleration [19] which has been fitted to the data [20].

2.3. *Isotopes*

Light element isotopic composition of CR as compared to that at sources can provide a detailed description of the CR propagation history for individual elements. In addition, the ratio of selected unstable to stable isotopes of elements of the same origin is directly related to the CR confinement time in the galaxy.

Isotope separation in a wide energy range is an experimental challenge. The measurements for H and He isotopes have been obtained using magnetic spectrometers. For heavier elements, most of the measurements come from space-borne spectrometers which are only able to perform isotope separation in a very limited energy range.

Deuterium to proton ratio has been measured in the range 0.2 GeV/n $\lesssim$ E $\lesssim$ 0.8 GeV/n by BESS [21] and AMS [22], whereas ^{3}He to ^{4}He ration has been measured in the range 0.1 GeV/n $\lesssim$ E $\lesssim$ 3.4 GeV/n by IMAX [23], BESS [21] and SMILI [24] spectrometers. Both measurements are dominated by statistical errors amounting 5% to 10%.

Figure 3 shows these measurements together with the expectations of the standard leaky-box calculation [23].

Among all β-radioactive secondary nuclei in CR, ^{10}Be is the lightest experimentally resolved isotope having a half-life comparable with the confinement time of CR in the galaxy ($t_{1/2}(^{10}\text{Be}) = 1.51$ Myr). Present measurements of the ratio

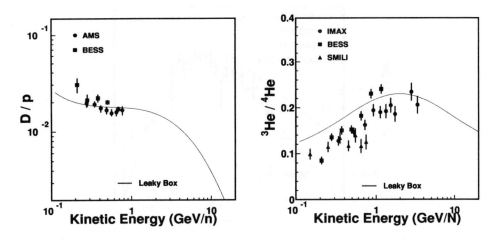

Fig. 3. Measurements of deuteron over proton and ^{3}He over ^{4}He ratios together with the expectations for a propagation model as described in the text.

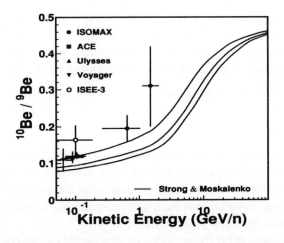

Fig. 4. Measurements of ^{10}Be over ^{9}Be ratio together with the expectations for a propagation model as described in the text.

of ^{10}Be to the stable ^{9}Be have been performed using space-borne spectrometers (ACE [25], ISEE-3 [26], Ulysses [27], Voyager [28]) for energies $\lesssim$ 100 MeV/n and with the ISOMAX balloon-borne magnetic spectrometer for energies in the range 0.26 – 2 GeV/n [29,30]. While measurements in the low energy range have a precision of 10%, the limited statistics in the ISOMAX measurements can only provide a 20% – 30% accuracy at higher energies.

Figure 4 shows the present measurements compared to the expectations of a propagation model [19] for three different values of the galactic halo size [20]. The

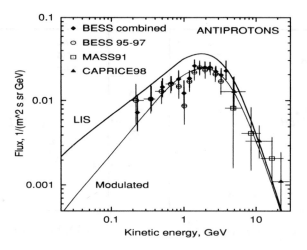

Fig. 5. Summary of recent antiproton flux measurements and the calculated local interstellar and modulated spectrum from the reference quoted in the text.

enhancement of the ^{10}Be/^{9}Be ratio at high energies is due to the Lorentz time expansion. The limiting value at high energies corresponds to the production ratio, whereas at lower energies the ratio decreases according to the combined effect of the finite confinement time of CR in the galaxy and the effective ^{10}Be half-life $\gamma \cdot t_{1/2}$.

2.4. Antiprotons

Antiprotons are the only hadronic antimatter detected in CR and are believed to be produced by interaction of primary CR with the interstellar medium. Distortions in the distinctive antiproton spectrum, related to its secondary origin, could arise from primary source contributions such as primordial black hole evaporation [31] or neutralino annihilation in the galactic halo [32].

Antiproton identification requires charge sign identification capabilities to fight the huge proton background ($\sim 10^4$) and electron-proton discrimination to reject the electron contamination ($\sim 10^2$). This task is reserved to magnetic spectrometers. The current measurements are dominated by BESS which has collected a total amount of about 2000 $\bar{p}$ in the range from 0.18 to 4.2 GeV after several flights [33,34,35,36]. Above this energy, MASS91 [37] and CAPRICE98 [38] have extended the antiproton measurements up to 50 GeV after collecting a total amount of about 50 antiprotons.

After the first observations of antiprotons in CR [39,40], which implied a large yield at low energies as expected from exotic origin, present measurements agree with the improved theoretical predictions for a pure secondary origin as shown in figure 5 coming from I.V. Moskalenko et al. [41].

3. Future Experiments

Most interesting measurements to be performed for understanding CR propagation rely on precise measurements at higher energies for the dominant hadronic CR and on good element and isotopic identification of less abundant species.

In order to achieve high precision measurements several limitations have to be overcome. Balloon-borne experiments are subject to short exposure times ($\sim$ 1 day) which makes them insensitive to faint signals and thus limiting the covered energy range. Space-borne experiments suffer from limited identification capabilities and small geometrical acceptance which, in turn, produces similar effects as the balloon-borne experiments.

The balloon-borne community has launched the Long Duration Balloon Project (LDB) and the Ultra-LDB (ULDB) programs which are intended to extend the current flights up to 100 days [42]. Despite the recent failure of two ULDB test balloons which resulted in a likely 2 year delay in the program, several LDB flights have been successfully operated. The JACEE collaboration has performed five 9-15 day antarctic circumpolar flights [43] and, more recently, the Advanced Thin Ionization Calorimeter (ATIC) has performed a 16-day flight [44]. In the near future, the BESS collaboration is preparing a 2-3 week antarctic flight to precisely measure the antiproton flux in the low energy range [45]. In the northern hemisphere, the Transition Radiation Array for Cosmic Energetic Radiation (TRACER) will perform a 2-week arctic circumpolar flight this year [46], and the new proposed experiment (Polar BEAR [47]), which involves a combination of large collecting area Kinematic Lightweight Energy Meter (KLEM [48]) with an ionization calorimeter, could be implemented for a LDB flight in 2004-2005.

The new generation of balloon-borne experiments is mostly devoted to the study of high energy nuclei up to energies of 10^{15} eV. This goal could be achieved after the exposure provided by ULDB flights or long series of LDB. Anyway, some of the techniques developed for these experiments could be applied in large-acceptance long-duration satellite instruments.

Regarding the space-borne future experiments, the PAMELA magnetic spectrometer is expected to be launched in space in 2003 for a more than 2-year mission [49]. PAMELA experimental program is devoted to the study of antiprotons and positrons. The AMS experiment [50], to be placed in the International Space Station (ISS) in 2005, is the upgraded version of the 1998 Shuttle flight magnetic spectrometer and, thanks to its large geometrical acceptance and long duration (3 years), will perform precise studies of hadrons, leptons and γ-rays. Finally, the Advanced Cosmic-ray Composition Experiment (ACCESS [51]) has been proposed for a long duration (3 years) exposure, either attached to the ISS or as a free-flyer satellite. The ACCESS experiment, composed of subdetectors based on the experience of CRIS/ACE, TRACER and ATIC, is designed to measure the abundances of all nuclei species in cosmic rays and the spectra of individual elements up to Z=28.

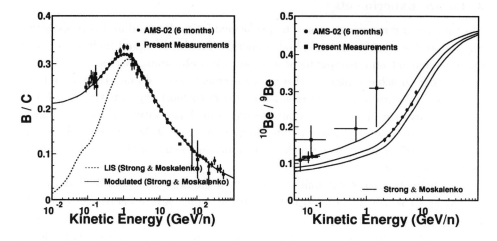

Fig. 6. AMS expected performances for B/C and ^{10}Be/^{9}Be ratios after 6 months of data taking compared to the present measurements. The ratios have been simulated according to the expectations for a propagation model described in the text.

4. Expected Performances

In order to illustrate the capabilities of near future large-acceptance long-duration experiments we will just outline the expected performances of some of them.

The BESS-polar long duration flight will collect after 20 days more than 10^3 antiprotons in the energy range from 0.1 GeV to 4.2 GeV [45].

The PAMELA experiment will be able to identify and measure antiprotons up to 190 GeV and after two years will collect more than 2×10^4 antiprotons [49].

The AMS detector will be able to collect and precisely measure enormous amount of data at high energy. AMS will identify and measure the energy of hydrogen and light nuclei for $E \lesssim 1$ TeV. After three years, AMS will collect 10^8 H with energies above 100 GeV and 10^7 He with energies above 100 GeV/n. Regarding the light nuclei above helium, AMS will collect $10^5 (10^4)$ carbon(boron) nuclei with energies above 100 GeV/n. Light isotope separation will be achievable up to 10 GeV/n and, after 3 years, AMS will identify 10^8 D, 10^8 ^{3}He and 10^5 ^{10}Be.

The simulated B/C and ^{10}Be/^{9}Be ratios achievable after a period of 6 months of data taking (one sixth of the total mission duration) are shown in figure 6 together with the present measurements and the expectations of a propagation model [19,20].

In addition, AMS is able to identify antiprotons up to 300 GeV and, after three years, is expected to collect 10^6 antiprotons.

Finally, ACCESS after a 3-year exposure should be able to measure the energy spectra of individual elements from hydrogen to nickel for $E \lesssim 10^{15}$ and the elemental abundances for all elements heavier than iron.

5. Conclusions

Precise knowledge of the hadronic component of cosmic rays is needed to describe the cosmic ray production, acceleration and propagation mechanisms in our galaxy. Propagation models, which can predict the backgrounds for rare signal searches in cosmic rays, have to be tuned using a set of precise measurements in a wide energy range. Present measurements, suffering from limitations coming from short exposure time and intrinsic instrumental limitations, have provided accurate results in a restricted energy range for several key quantities. Near future experiments have been designed to overcome the previous limitations using sensitive large-acceptance long-duration instruments. The results obtained with these new experiments will improve dramatically our current knowledge of cosmic rays.

Acknowledgements

I would like to thank Prof. Bruna Bertucci for her patience in waiting for this contribution to be written.

References

1. T. Sanuki et al., *Astrophys. J.* **545** 1135 (2000).
2. W. Menn et al., *Astrophys. J.* **533** 281 (2000).
3. M. Boezio et al., *Astrophys. J.* **518** 457 (1999).
4. J. Alcaraz et al., *Phys. Lett.* **B490** 27 (2000).
5. J. Alcaraz et al., *Phys. Lett.* **B494** 193 (2000).
6. M.J. Ryan et al., *Phys. Rev. Lett.* **28** 985 (1972).
7. J. Buckley et al., *Astrophys. J.* **429** 736 (1994).
8. K. Asakimori et al., *Astrophys. J.* **502** 278 (1998).
9. A.V. Apanasenko et al., *Astropart. Phys.* **16** 13 (2001).
10. T.K. Gaisser, *Astropart. Phys.* **16** 285 (2002).
11. T.K. Gaisser, T. Stanev, *Phys. Rev.* **D57** 1977 (1998).
12. T.K. Gaisser et al., *Proc. 27th ICRC* p.1643 (2001).
13. J.J. Engelmann et al., *Astron. Astrophys.* **233** 96 (1990).
14. D. Müller et al., *Astrophys. J.* **374** 356 (1991).
15. S.P. Swordy et al., *Astrophys. J.* **349** 625 (1990).
16. R. Dwyer, P. Meyer, *Astrophys. J.* **322** 981 (1987).
17. K.E. Krombel, M.E. Wiedenbeck, *Astrophys. J.* **328** 940 (1988).
18. A.J. Davis et al., *AIP Conf. Proc.* **528** 421 (2000).
19. A.W. Strong, I.V. Moskalenko, *Astrophys. J.* **509** 212 (1998).
20. A.W. Strong, I.V. Moskalenko, *Adv. Space Res.* **27** 717 (2001).
21. E.S. Seo et al., *Proc. 25th ICRC* **3** 373 (1997).
22. G. Lamanna et al., *Proc. 27th ICRC* 1614 (2001).
23. A.J. Davis et al., *Proc. 24th ICRC* **2** 622 (1995).
24. J.J. Beatty et al., *Astrophys. J.* **413** 268 (1993).
25. W.R. Binns et al., *Proc. 26th ICRC* **3** 21 (1999).
26. M.E. Wiedenbeck, *Proc. 19th ICRC* **2** 84 (1985).
27. J.J. Connell, *Astrophys. J.* **501** L59 (1998).
28. A. Lukasiak et al., *Proc. 26th ICRC* **3** 41 (1999).
29. T. Hams et al., *Proc. 27th ICRC* 1655 (2001).

30. G.A. de Nolfo et al., *Proc. 27th ICRC* 1659 (2001).
31. K. Maki et al., *Phys. Rev. Lett.* **76** 3474 (1996).
32. G. Jungman et al., *Phys. Rep.* **267** 195 (1996).
33. K. Yoshimura et al., *Phys. Rev. Lett.* **75** (1995) 3792.
34. S. Orito et al., *Phys. Rev. Lett.* **84** 1078 (2000).
35. T. Maneo et al., *Astropart. Phys.* **16** 121 (2001).
36. Y. Asaoka et al., *Phys. Rev. Lett.* **88** 051101 (2002).
37. G. Basini et al., *Proc. 26th ICRC* **3** 77 (1999).
38. M. Boezio et al., *Astrophys. J.* **561** 787 (2001).
39. R.L. Golden et al., *Phys. Rev. Lett.* **43** 1196(1979) .
40. A. Buffington et al., *Astrophys. J.* **248** 1179(1981) .
41. I.V. Moskalenko et al., *Astrophys J.* **565** 280 (2002).
42. http://www.wff.nasa.gov/~uldb
43. R.J. Wilkes et al., *Proc. 24th ICRC* **3** (1995) 615.
44. J.P. Wefel, *Proc. 27th ICRC* (2001).
45. A. Yamamoto et al., *Proc. 27th ICRC* 2135 (2001).
46. F. Gahbauer et al., *Proc. 27th ICRC* 1612 (2001).
47. G. Bashindzhagyan et al., *Proc. 27th ICRC* 2147 (2001).
48. J. Adams et al., *Proc. STAIF* **504** 175 (2000).
49. P. Spillantini, *Proc. 27th ICRC* 2215 (2001).
50. http://ams.cern.ch/AMS
51. http://wwwmipd.gsfc.nasa.gov/access/access.htm

REVIEW OF PRECISION MEASUREMENTS OF HIGH ENERGY ELECTRONS

BRUNA BERTUCCI

Università di Perugia e Sezione INFN di Perugia
Via Pascoli, Perugia, 06124 Italy

An accurate measurement of the intensity and energy spectra of Cosmic Ray electrons and positrons represents a major experimental challenge. Long exposure times and excellent particle identification capabilities are needed in order to cope with the low intensity of the electron and positron fluxes and the overwhelming background from protons and nuclei in cosmic rays. The motivations for such an experimental effort will be briefly discussed and the most recent results revieweved together with the perspectives of future experiments.

Keywords: Cosmic Ray; Electrons; Positrons

1. Introduction

Electrons[a] represent only a small fraction, O(%), of the Cosmic Rays reaching the earth's atmosphere, nevertheless the astrophysical importance of their flux measurement and charge composition is fully recognized and has triggered a continuous experimental effort during the last 40 years.

As a matter of fact, the electron flux at the top of the atmosphere is dominated by the negative component (90% @ 1 GeV) pointing to a primary origin for the bulk of e^-, directly injected at sources, and to a secondary origin of e^+, resulting from nuclear collisions in the ISM.

Different constraints on propagation models are then carried by separate $e^\pm$ measurements. The expected positron flux at earth heavily depends on the characteristics of the nuclear CR component and its propagation history, dominated by spallation and energy losses in hadronic interactions with the residual matter of the ISM. Electromagnetic interactions are fully responsible for the dynamics of electron propagation from sources to the earth: interactions with photons and magnetic fields of the ISM lead to distortions of the original e^- energy spectrum which can be related to the space/time scale of the journey from acceleration sites to the earth.

---

[a]in the following we will generically indicate as electrons both e^- and e^+ whenever the charge sign is not explicitely stated.

1.1. The e^- energy spectrum

Dynamics of electron propagation from acceleration sites to us is dominated by the interaction with matter (bremsstrahlung and ionization) at energies below few GeVs and by the interaction with magnetics fields (synchrotron emission) and background photons (inverse compton scattering) above.

In fact, much information about electrons fluxes at sources and in the galaxy can be extracted from the indirect measurement of their e.m. radiation over a broad range of energy, going from the radio up to the γ-ray emission. Multiwavelength analysis of e.m. emission at supernovae remnants have identified electron sources, capable of accelerating them up to highest energies [1], measurements of low energy γ rays (E_γ <100 MeV) produced in electron bremsstrahlung and IC scattering are used to infer the electron fluxes in the 10-100 MeV energy range, radio emission spectra in our galaxy have been translated into energy spectra of synchrotron radiating electrons[2].

However only direct measurement of the electron fluxes over a wide range of energies can really give us a solid basis to test different hypothesis on source distributions, injection spectra and propagation models, the basic question always being whether the flux measured near earth is representative of the *average* galactic population or depends on local conditions.

At low energies, below O(10 GeV), electron fluxes near earth are not even representative of the local interstellar spectrum due to geomagnetic and solar modulation effects. Direct measurements are relevant for studies of the earth's magnetosphere [3,4,5] and of the interplanetary medium, charge effects in solar modulation has been studied just based upon electron spectra [6,7].

At the other extreme of the spectrum, O(TeV), the electron energies are such that the electron cooling time in the ISM[b] is $O(10^5)$ years, relatively short with respect to a conventional CR confinement time of $O(10^7)$ years. What we can in general expect is that the energy spectrum depends from the propagation conditions in the ISM through regions with a size set by the diffusion lenght $\lambda(E) \simeq (\frac{2D}{bE})^{\frac{1}{2}}$. For typical values of the diffusion coefficient D$\approx 10^{28}\ cm^2\ s^{-1}$, this sets an limit of 1Kpc on the distance of the sources. Again, we probe a relatively *small* region of our galaxy, and possibly *young* sources. Any model trying to predict such region of the spectrum must follow a discrete source approach rather than assuming an uniform source distribution as in conventional propagation models [8].

---

[b]We remind that the rate of energy loss for electrons above few GeV in the IS medium is due to synchrotron radiation and IC scattering, and can be expressed as:

$$\frac{dE}{dt} = -b \cdot (w_{ph} + w_B) \cdot E^2 = -k \cdot E^2\ [GeV/s] \qquad (1)$$

where $w_{ph}, w_B [eV/cm^{-3}]$ represent the energy densities of photons and magnetic fields in the ISM and $b \approx 10^{-16}$. The electron energy can be then written as $E(t) = E_0/(1 + k \cdot E_0 \cdot t)$. The cooling time needed to reach an energy E starting from $E_0 >> E$ is therefore $\tau = \frac{1}{kE}$ and since $(w_{ph} + w_B) \approx 1 eV/cm^3$ we get $\tau \approx 3 \cdot 10^8$ years /E(GeV)

At intermediate energies, what we can qualitatively expect is that the injection spectrum at source, represented as a power law $E^{-\alpha}$, is steepened by the continuous energy losses [9]. The actual prediction on the observable spectral index $\alpha' = \alpha + k$ varies on the different assumptions used in the solution of the diffusion-loss equation, the k value reaching the unity in a steady state were the electrons' life time is longer than the escape time from a confinement volume, while somewhat lower values are predicted for energy dependent diffusive models.

Different regimes in the spectral index of the electrons' flux measured at earth can be therefore expected reflecting different *boundaries* in the source distribution (galactic halo → galactic disk → nearby discrete sources) and/or propagation conditions. An accurate experimental determination of the electron energy spectrum over a wide energy interval is mandatory in order to better constraint and disentangle the proposed propagation models.

1.2. *The e^+ energy spectrum*

The bulk of the cosmic ray positrons observed at energies above $O(10^8$ eV$)$ are thought to be of secondary origin, resulting from nuclear interaction of cosmic rays, through the $\pi \to \mu \to e$ decay chain. Therefore many ingredients enter into the calculation of the expected positron abundance at earth: the composition, energy spectra and propagation history of nuclear CR interstellar fluxes, the matter distribution in the ISM, the nuclear cross sections and the description of the fragmentation process. Just from this fact can be easily guessed the difficulty of getting any robust prediction on the e^+ flux, but also the relevance of such a measurement as a test of propagation models in a way complementar to the direct measurement of CR nuclear abundances and isotopic composition. In this context, positrons, as $\bar{p}$ and γ's from π^0 decay, represent not only an over-constraint to different cosmic models but could also represent a potential probe for new physics, whenever an *unexpected* abundance would show up pointing to an exotic primary origin.

Large fluctuations in experimental data have triggered since many years a large theoretical effort in order to assess or discard a possible primary component in the observed e^+ flux. Above 10 GeV, where solar modulation effects start to be negligible, a positron fraction of a few % - monotonically decreasing with energy - is predicted in conventional leaky-box [10] or diffusion models [11]. Up to mid 90's, the experimental results were indicating a positron fraction as large as 20% and slowly increasing with energy[12,37,38,41]. This evidence supported theories were a large contribution to the positron flux was coming from other astrophysical processes, as mini-black hole evaporation [13], $e^{\pm}$ pair production in e.m. cascades initiated by high energy γ's near discrete sources [14,15], hadronic interaction in giant molecular clouds [16], or from more exotic sources, as anti-matter of extra galactic origin [17] and annihilation of Weakly Interacting Massive Particles (WIMPs) in the galactic halo [18,19,20]. More recent measurements [44,46,47,48] indicate a value of the positron flux sensibly lower that in the past, pointing to a smaller contribution - if any - from unconventional sources. As discussed in [21], there are still features in the observed spectra which cannot be fully accounted by standard secondary production and are used to limit the parameter space in different models[22,23], however no definite conclusion can be drawn still and only more accurate experiments over a wider energy range could help in this concern.

2. The experimental techniques

The key issues in the direct measurement of CR electron fluxes are:

(1) the low intensity of the signal
(2) the high background from CR nuclear component: the proton to electron ratio goes from $\sim 10^2$ at energies around 10 GeV rising up to $10^{3 \div 4}$ at O(TeV). For positrons the same figures should be multiplied by at least a factor 10.
(3) the copious production of $e^{\pm}$ in the interactions of nuclear cosmic rays with the atmosphere: for atmospheric depths of 6 gr $\cdot$ cm^{-2} the secondary flux contributes with a 5(10)% to the $e^-(e^+)$ flux at energies ≈ 10 GeV.

A large acceptance and long exposure times, an excellent e/p separation capability, the operation on a stratospheric balloon or a space based facility are the key requirements which have guided the design of all experiments venturing on CR electrons' measurement during the last 30 years.

Severe constraints are set on the weight and dimensions of the detectors as well as on the exposure time by the carrier technology, this gives rise to quite different experimental approaches whether the focus is on the measurement of the all electron spectrum up to the highest energies or on the separate measurement of e^- and e^+ components.

A rejection factor against protons of $O(10^5)$ is good enough to extend the all electron measurement to the TeV range, the main problem being to have a large enough acceptance to increase the statistics in a region were the flux is extremely low ($\sim 10^{-7 \div 8}$ part/GeV$\cdot$m$^2\cdot$s). A similar rejection factor is barely sufficient instead to get the e^+ measurement up to the 100 GeV range, with the further need of an accurate measurement of the charge sign up to that energy to reject the e^-.

In Tab.1 are listed the experiments which have played a major role in the measurement of electron fluxes and the energy interval covered by their results. The two sections of the table distinguish detectors without charge sign measurement capability (top) from magnetic spectrometers (bottom). The energy interval covered by each measurement is also reported, values in parenthesis correspond to the upper energy for positrons. We should advance a *caveat*: the upper value of the interval represents the highest energy recorded for a given experiment, but quite lower median energy values can characterize the corresponding energy bins in the flux determination. As an example, the TS93 experiment[43] measures positrons up to 63 GeV, but the last bin in the flux determination is based on 10 positrons over a 40 GeV energy interval. This corresponds to a 22 GeV median energy in flux de-

Table 1. Overview of experiments measuring CR electron/positron fluxes

Experiment	Year	Energy (GeV)	Measurement
Nishimura et al.[29]	1968-1976	30-1000	all electrons
Nishimura et al.[30]	1996-1998	30-3000	all electrons
Muller and Meyer[33]	1970	10-900	all electrons
Prince et al.[34]	1975	9-300	all electrons
Tang et al.[35]	1980	5-300	all electrons
Muller and Tang.[38]	1984	10-20	all electrons, e^+ fraction
BETS[31]	1997-1998	10-100	all electrons
Fanselow et al.[39]	1965-1966	0.05-14.3	e^-, e^+, e^+ fraction
Buffington et al.[12]	1972	4-50	e^-, e^+ e^+ fraction
Golden et al.[40,41]	1976	4.5-63.5	e^-, e^+, e^+ fraction
MASS[42]	1989	4-13	e^-, e^+, e^+ fraction
TS93[43]	1993	5-60	e^+ fraction
CAPRICE[44,45]	1994	0.46-43.6	e^-, e^+, e^+ fraction
CAPRICE[46]	1998	4-30	e^-, e^+, e^+ fraction
HEAT [47,48]	1994	5-100 (50)	e^-, e^+, e^+ fraction
HEAT [49,50]	1995	1-100 (50)	e^-, e^+, e^+ fraction
HEAT/PBAR[51]	2000	5-16.4	e^+ fraction
AMS[25]	1998	0.2-40(3)	e^-, e^+, e^+ fraction

termination, to be compared with a median of 34 GeV for the HEAT[50] experiment, which has a lower maximum energy.

Compact shower detectors, with different technical implementations, have reached by now the highest energies in the all electron spectrum measurement. In this approach, the geometry of the instrument allows for a large field of view which enhances the acceptance. Emulsion chambers[29,30], made of a sandwich of Pb or W interleaved with emulsion plates and X-ray films, allow a very clear separation of electron induced cascades from hadronic or photon showers, leading to an estimated p background < 10% in the signal. Main drawbacks of this technique are the pile-up of events, which practically prevents continuous exposure during long periods, and the absence of a trigger: no time information is associated to the recorded events and no correlation with a potential nearby source can be therefore extracted. Modern imaging calorimeters [31], made of scintillating fibers in lead, point to achieve the same accuracy in calorimetric separation of e.m and hadronic showers with added trigger capability. However, a common problem of such kind of experiments is the lack of calibration of the detector response at energies above a few hundred GeVs. The estimated energy resolution and proton discrimination heavily rely on simulation of the detector guided by calibration at much lower energies, limited cross checks, if any, can be effectively done on data themselves.

In a more general approach also the relativistic velocity of electrons has been exploited for signal selection by means of a large variety of techniques: Time of Flight (TOF) hodoscopes, Transition Radiation and Cerenkov light detectors have been employed in conjunction with shower counters and magnetic spectrometers in order to get redundancy in particle identification. While increasing the overall rejection power against protons, this allows a direct cross-check of the efficiencies for signal and background in each step of the analysis, minimizing the influence of the detector simulation in the result. With this approach in mind, most of the experiments listed in the table are from a few collaborations which have evolved along many years with continuous upgrades of their balloon borne instruments. As an example, results from [40–46] make use of the same superconducting magnet [40] equipped in subsequent flights with different tracking chambers (Micro Wire Proportional Chambers[40,42] + Drift Chambers[43–46]), calorimeters (Pb+Scintillators[40], Brass+streamer tubes[42], Si-W[43–46]), Cerenkov Detectors (Gaseous Threshold[40,42], solid Ring Imaging Cerenkov (RICH)[45], Gaseous RICH[46]),Transition Radiation Detector[43]. A quite different position is taken by AMS in this concern, a magnetic spectrometer conceived for a three years' mission in space on board of the International Space Station. Its actual results on electrons[25] are coming from data gathered on shuttle during a test flight in 1998, where a limited redundancy in particle identification based on velocity was granted by the Time of Flight system and a Aereogel Threshold Cerenkov (ATC) counter. This limits a clean positron identification to few GeVs, much lower than in most measurements, while a wider energy interval is covered for e^- thanks to the proton rejection based on charge sign determination.

3. The e^- spectrum

In Fig.1 measurements of the all electrons flux as a function of energy are reported, multiplied by an energy dependent factor - E^3 - in order to better appreciate different results over 3 orders of magnitude in flux intensity. Open symbols have

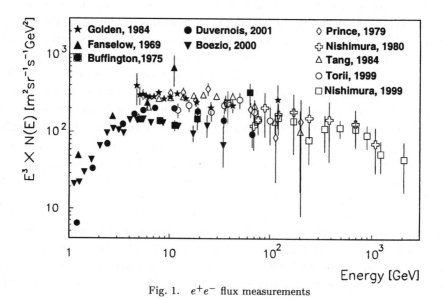

Fig. 1. e^+e^- flux measurements

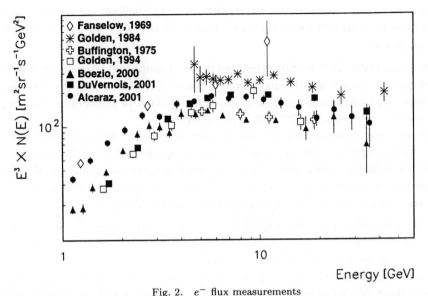

Fig. 2. e^- flux measurements

been choosen for experiments with no charge measurement capability, grey and black filled symbols refer to older and most recent measurements with magnetic spectrometers.

As a general tendence, results from experiments with no charge sign measurement point towards higher values of the electron flux. This can be clearly seen in Fig.1 where the open symbols lie systematically above the black filled ones. An underestimation of the proton background could be fairly assumed as a relevant part of this effect, since little redundancy in electron identification is available in these experiments.

However, even considering only direct measurement of the e^- spectrum from magnetic spectrometers, the various results can differ by a factor ≈ 2, as can be better seen in Fig.2. The differences are less pronounced, but still around a 40-50%, among most recent results which benefit of more advanced analysis techniques. They are represented in Fig.2 by grey and black filled symbols.

It is hard to point out a single cause at the basis of this dispersion, which results from the concurrency of detector related systematic effects (acceptance, resolution, background rejection) and different procedures in the evaluation of atmospheric corrections. As an example of detector related effects, completely different sources induce a systematic uncertainty [c] on the flux normalization of O(10%) both for the CAPRICE (Boezio, 2000 [45]) and the HEAT (DuVernois, 2001 [50]) results. In CAPRICE, a 8% uncertainty on the tracking efficiency directly reflects on the flux normalization, in HEAT a few percent uncertainty on the energy scale[48] translates in a systematic error of 10% due to the steep power law of the electron spectrum.

---

[c]not included in the error bars of the data in Fig.2

The presence of some residual atmosphere on top of balloon borne experiments leads to a twofold correction on the fluxes measured at the payload in order to get the galactic ones:

(1) the subtraction of the e⁻ flux generated in the atmosphere by the cosmic nuclear component. This is not the main issue in the e⁻ measurement above few GeVs, where this background represents only a few percent of the total e⁻ flux with a relative uncertainty of 10-20% in its estimate.
(2) the rescaling of the measured energies to compensate the bremsstrahlung losses of cosmic electrons in the atmosphere: an energy shift of $\approx$ 5-10% corresponds to typical floating altitudes. This is somewhat a more delicate issue, since the uncertainty in the energy scale is amplified by the spectral index for the flux normalization. Quoted experimental errors due to this are $\approx$ 5%.

Quite different techniques are adopted by the various experiments to evaluate both atmosperic related corrections. Extensive checks of the used models for background subtraction are performed against atmospheric growth curves of $e^{\pm}$ or muon fluxes and quite conservative errors are quoted for the uncertainty on the energy losses. However, large discrepancies are found in direct comparisons of the various calculations and data points of a given experiment could move well outside the quoted errors if different corrections were to be applied. With the given experimental situation, it is not really from the electrons spectra that a stringent test to various propagation models can be performed. While an overall indication of a spectral index > 3 appears from most data sets, quantitative analysis both on the shape and intensity are practically forbidden by the small statistics at the highest energies and the large spread of measurements.

4. The e^+ spectrum and $e^+/(e^+ + e^-)$

In Fig.3 are shown the results on positron fluxes as a function of energy, scaled by E^3. It can be immediately noticed that most of the measurements lie in the low energy part of the spectrum and are affected by large errors due to the severe experimental requirements to extend the measurement above several GeVs. The main limiting factors come from the low intensity of the signal and the presence of various sources of backgrounds, which constrain different aspects of the measurement:

(1) cosmic protons: the p/e^+ flux ratio is $O(10^4)$ at 10 GeV. No distinction can be done based on the charge sign, moreover particle identification based on velocity becomes more and more difficult at increasing energies. It is hard to obtain a reliable rejection by means of a single technique.
(2) cosmic electrons: the e^-/e^+ ratio is $O(10)$ at 10 GeV. The only possible distinction[d] is on the basis of the charge sign: is then mandatory the use of a magnetic

---
[d]Earlier experiments[36,37,38] used the asymmetry in the geomagnetic cutoff rigidity to to separate $e^{\pm}$. Only the relative fration can be measured in this way, but large uncertainties affect the measurement

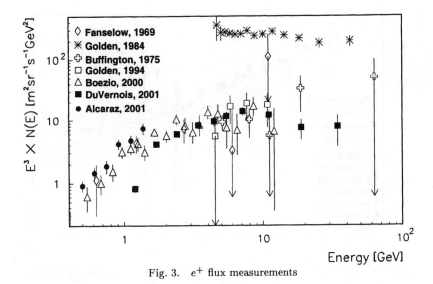

Fig. 3. e^+ flux measurements

spectrometer with a large tracking volume and a strong magnetic field in order to keep low the charge confusion probability event at high rigidities.

(3) atmospheric e^+: atmospheric production of electrons is charge sign dependent enhancing the secondary positron component mostly at low energies. The relative importance of this component should in principle decrease with increasing energy, however at 10 GeV it still represents a 20 − 30% of the positron flux. Since this background is virtually indistinguishable from the signal, it can only be subracted by means of its theoretical estimates, with large systematic uncertainties associated in this procedure.

As an example, the CAPRICE (Boezio, 2000 [45]) measurement ends at 10 GeV due to proton contamination. Above 5 GV their RICH detector is not capable of e/p separation, which is therefore done only by the calorimeter, leading to a 50% systematic uncertainty on the flux. In HEAT (Barwick, 1998 [48]) the limit in energy is due the spectrometer capability of measuring the charge sign up to 50 GV.

With the large errors associated to the positron measurements is quite uneasy to perform any comparison with propagation models. However, part of the experimental uncertainties common to the e^+ and e^- measurements cancel out in the positron to electron flux ratio, $e^+/(e^+ + e^-)$, which is therefore widely used in the comparison with the theoretical curves.

In Fig.4 the positron fraction measurements are shown as a function of the energy, older measurements are reported with emtpy symbols. The solid and dashed curves superimposed to data are the predictions for a purely secondary origin of cosmic ray positrons in a framework of the leaky-box model[10] and a diffusive pro-

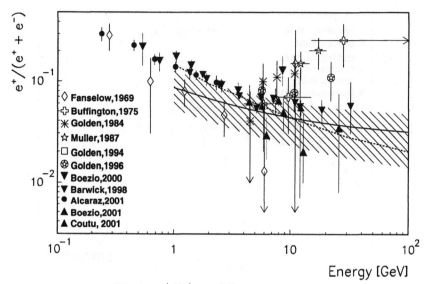

Fig. 4. $e^+/(e^+ + e^-)$ flux measurements

pagation model[11] respectively. The dashed area corresponds to the uncertainty on the leaky-box prediction, but a similar band is appropriate also for the other model.

As mentioned in Sec.1.2, early measurements have indicated for a long while a positron fraction increasing with energy and in clear excess with the theoretical estimates. This has been at the origin of many speculations about primary sources of positrons. As of today, measurements from HEAT (Barwick, 1998[48], Coutu, 2001[51]) and CAPRICE (Boezio, 2000[45], Boezio, 2001[46]) have a much lower normalization, generally compatible with a secondary origin. Nevertheless the debate on the existence of a primary component is still alive mainly due to the following arguments:

(1) there is some indication of a structure around 7 GeV in the energy dependence of the positron fraction as measured by HEAT. This feature is not comfortably accounted by the smooth energy dependence expected for products of the nuclear collisions in the interstellar medium.
(2) in the quest of anti-matter as a signature of dark matter annihilation, the positron signal is a promising channel. The main advantages are a relatively small e^+ flux from ordinary sources and the smaller dependence on large scale structure of the galaxy, since the energy losses reduce the possible source distances to few Kpcs.

Only new measurements over a larger energy range and better statistics will be able to better constraint propagation models and give a clear signal of new physics.

Table 2. Overview of future experiments measuring CR electron/positron fluxes

Experiment	Location	Year	Duration	Geometrical factor ($m^2 sr$)	Energy (GeV)	e^+ E>10 GeV	e^- E>100 GeV
ATIC	Balloon	2000	16 days	0.25	10-1000	–	$10^2 \div 10^3$
PAMELA	Satellite	2003	3 years	$2.7 \cdot 10^{-3}$	0.05-300	$O(10^4)$	$10^2 \div 10^3$
AMS	Space Station	2004	3 years	0.045	1-300 GeV	$O(10^5)$	$10^3 \div 10^4$
CALET	Space Station	2007	3 years	$0.5 \div 1$	$10-10^4$	–	$10^4 \div 10^5$

5. Perspectives for the future

A new generation of experiments is currently in preparation for long exposures in space (PAMELA[64], AMS02[65], CALET[66]), while analysis of results from long duration balloon flight (ATIC[67]) is in progress. In Tab.2 the main characteristics of the experiments are reviewed together with the foreseen statistics that they will collect at high energies for the e^+, e^- channels. For AMS02, the geometrical factor for e^+ is quoted, limited by the calorimetric acceptance. It includes also analysis cuts [68]. A striking improvement in the accuracy of electrons flux and composition determination is expected up to TeV energies, confirming the entrance of experimental astrophysics in the domain of the precision measurements.

References

1. G. E. Allen, R. Petre and E. V. Gotthelf, *Astrophys. J.* **558**, 739 (2001).
2. J. M. Rockstroh and W. R. Webber, *Astrophys. J.* **224**, 677 (1978); W. R. Webber, G. A. Simpson and H.V. Cane, *Astrophys. J.* **236**, 448 (1980);
3. E. Fiandrini, G. Esposito, contribution to this workshop
4. V. Mikhailov, contribution to this workshop
5. D. Heynderyxx, contribution to this workshop
6. P. Evenson et al, *Journ. Geoph. Res.* **100**, 7873 (1995).
7. J. M. Clem, D. P. Clemens et al, *Astrophys. J.* **464**, 507 (1996).
8. A. M. Atoyan, F. A. Aharonian and H. J. Volk , *Phys. Rev.* **D52**, 3265 (1995).
9. for a qualitative discussion on the subject see for instance M.S. Longair, *High energy Astrophysics*, Vol. II, chap.19
10. R. J. Protheroe Astrophys. Journ. 254 , 391 (1982).
11. I.V. Moskalenko and A. W. Strong , *Astrophys. J.* **493**, 694 (1998).
12. A. Buffington, C. D. Orth and G. F. Smoot, *Astrophys. J.* **199**, 669 (1975).
13. B. J. Carr, *Astrophys. J* **254**, 391 (1976);
14. A. Harding and R. Ramary, *Proc 24th Int. Cosmic Rays Conf. (Rome)* **4**, 1029 (1987);
15. F. A. Aharonian and A.M. Atoyan, *J Phys.G: Nucl Part. Phys.* **17**, 1969 (1991);
16. V. A. Dogiel and G.S. Sharov, *Astron. and Astrophys* **229**, 259 (1990);
17. R. L. Golden and S. A. Stephens, *Sp. Sc. Rev.* **46**, 3 (1987);
18. A.J. Tylka, *Phys. Rev. Lett* **75**, 3792 (1989);
19. M. S. Turner, F. Wilczek, *Phys. Rev.***D 42**, 1774 (1991);
20. M. Kamionkoswski, M. S. Turner, *Phys. Rev.***D 43**, 1774 (1991);
21. S. Coutu et al. Astropart. Phys. **11**, 429 (1999).
22. E. A. Baltz et al, ASTRO-PH/0109318
23. G. L. Kane, L. T. Tang and J. D. Wells, HEP-PH/0108138
24. V. L. Ginzburg and V. S. Ptuskin, *Rev. Mod. Phys.* **48**, 161 (1976).

25. J. Alcaraz et al, *Phys. Lett.* **B484**, 10 (2001).
26. J. A. Earl, *Phys. Rev. Lett.* **6**, 125 (1961).
27. P. Meyer and Vogt, *Phys. Rev. Lett.***6**, 193 (1961).
28. De Shong, R. H. Hildebrand and P.Meyer, *Phys. Rev. Lett.* **12**, 3 (1964).
29. J. Nishimura, *Astrophys. J.* **238**, 394 (1980).
30. J. Nishimura et al., *Adv. Space Res.* **26**, 1827 (2000).
31. Torii et al, *Adv. Space Res.* **26**, 1823 (2000).
32. Torii et al, *Nucl. Instr. and Meth.* **A452**, 81 (2000).
33. D.Muller et al, *Astrophys. J.* **186**, 841 (1973).
34. T. A. Prince, *Astrophys. J.* **227**, 676 (1979).
35. K. K. Tang, *Astrophys. J.* **278**, 881 (1984).
36. K. Anand et al, *Proc. 11th Int. Cosmic Ray Conf. (Budapest)* **1**, 235 (1969).
37. B. Agrinier, *Lett. Nuovo Cimento* **1**, 53 (1969);
38. D. Muller and K. K. Tang, *Astrophys. J.* **312**, 183 (1987).
39. J. L. Fanselow et al., *Astrophys. J.* **158**, 771 (1969).
40. R. L. Golden et al, *Astrophys. J.* **287**, 622 (1984).
41. R. L. Golden et al, *Astron. Astrophys.* **188**, 145 (1987).
42. R. L. Golden et al, *Astrophys. J.* **436**, 769 (1994).
43. R. L. Golden et al, *Astrophys. J.*, **457L**, 103 (1996).
44. G. Barbiellini et al, *Astron. Astrophys.* **309**, L15 (1996).
45. M. Boezio et al, *Astrophys. J.* **532**, 653 (2000).
46. M. Boezio et al, *AdSR* **26**, 669 (2001).
47. S. W. Barwick et al, *Phys. Rev Lett.* **75**, 390 (1995).
48. S. W. Barwick et al, *Astrophys. J.* **498**, 779 (1998).
49. S. W. Barwick et al, *Astrophys. J.* **482L**, 191 (1997).
50. M. A. DuVernois et al, *Astrophys. J.* **559**, 296 (2001).
51. S. Coutu et al, *Proc. 27th Int. Cosmic Ray Conf. (Hamburg)*, p.1687 (2001).
52. S. W. Barwick et al, *Nucl. Instr. and Meth.* **400**, 34 (1997).
53. Feng et al, ASTRO-PH/0008115
54. Milne et al, ASTRO-PH/0106157
55. R.Sma and E.S.Seo, *Adv. Space Res.* **26**, 1859 (2001).
56. C. D. Orth and A. Buffington *Astrophys. J.* **206**, 312 (1976).
57. A. W. Strong and I.V. Moskalenko, *Adv. Space Res.* **26** (2001).
58. F. L. Gualandris and A. W. Strong, *Astron. Astrophys* **140**, 357 (1984).
59. Daniel and Stephens, *Phys. Rev. Lett* **17**, 935 (1966).
60. R. R. Daniel and S. A. Stephens, *Phys. Rev. Lett.*,**15**, 769 (1965).
61. R. R. Daniel and S. A. Stephens, *Rev. Geophys. Space Phys.* ,**12**, 233(1974).
62. S. A. Stephens, *Proc. 17th Int. Cosmic Ray Conf. (Paris)*,**4**, 282 (1981).
63. M. Boezio et al, *Phys. Rev. Lett* **82**, 4757 (1999).
64. P. Spillantini, *Proc. 27th Int. Cosmic Ray Conf. (Hamburg)*, p 2215 (2001)
65. R.Battiston and references therein, contribution to this workshop, p.1
66. Torii, *Proc. 17th Int. Cosmic Ray Conf. (Hamburg)* p.2227 (2001).
67. J.Chang and W.K.H. Schmidt, *Proc. 27th Int. Cosmic Ray Conf. (Hamburg)*, p 2115 (2001).
68. V.Choutko et al., contribution to this workshop.

A MONTE CARLO SIMULATION OF THE COSMIC RAYS INTERACTIONS WITH THE EARTH'S ATMOSPHERE

P.ZUCCON*

Università degli Studi di Perugia and Sezione INFN
Via Pascoli, 06123 Perugia, Italy.

Substantial fluxes of protons and leptons with energies below the geomagnetic cutoff have been measured by the AMS experiment at altitudes of 370-390 Km, in the latitude interval ±51.7°. The production mechanisms of the observed trapped fluxes are investigated in detail by means of the FLUKA Monte Carlo simulation code. All known processes involved in the interaction of the cosmic protons with the atmosphere (detailed descriptions of the magnetic field and atmospheric density, as well as the electromagnetic and nuclear interaction processes) are included in the simulation. The results are presented and compared with the experimental data, indicating good agreement with the observed fluxes. The impact of secondary proton flux on particle production in atmosphere is briefly discussed.

Keywords: Cosmic rays, neutrinos, AMS

1. Introduction

Cosmic rays approaching the Earth interact with the atmosphere resulting in a substantial flux of secondary particles. The knowledge of composition, intensity and energy spectra of these particles is of considerable interest, e.g. for the evaluation of background radiation for satellites and the estimate of the atmospheric neutrino production for neutrino oscillation experiments [1].

The AMS measurements in near earth orbit [2,3] have allowed, for the first time, to gather accurate information on the intensity, energy spectra and geographical origin of charged particle fluxes at energies below the geomagnetic cutoff over a wide range of latitudes and at almost all longitudes. The under cutoff component of proton fluxes at equatorial latitudes has revealed an unexpected intensity of up to 50% of the primary proton flux, a positron to electron flux ratio has been found in the undercutoff component which largely exceeds the cosmic one.

A robust interpretation of these and many other characteristics of the undercutoff fluxes in terms of secondary particles produced in atmosphere requires an accurate description of both the interaction processes at their origin and of the geomagnetic field effects.

---

*E-mail: paolo.zuccon@pg.infn.it

In this work, we report results from a fully 3D Monte Carlo simulation based on FLUKA 2000 [5] for the description of cosmic ray interactions with the atmosphere. The key features of our analysis are an efficient generation technique for the incoming proton flux and a true microscopic, theory driven treatment of the interaction processes opposite to empirical parametrization of accelerator data. As a first attempt the contribution of He and the heavier nuclei, representing $(\approx 9\%)^6$ of the all nuclei cosmic flux, is neglected.

2. The model

An isotropic flux of protons is uniformly generated on a geocentric spherical surface with a radius of 1.07 Earth radii ($\sim 500\,Km$ a.s.l.) in the kinetic energy range $0.1-170\,GeV$.

We took the functional form suggested in [7] to describe the proton energy spectrum, the spectral index and the solar modulation parameter are extracted from a fit to the AMS data [8].

The magnetic field in the Earth's proximity includes two components: the Earth's magnetic field, calculated using a 10 harmonics IGRF [9] implementation, and the external magnetic field, calculated using the Tsyganenko Model[a] [10]. To account for the geomagnetic effects, for each primary proton we back-trace an antiproton of the same energy until one of the following conditions is satisfied:

(1) the particle reaches the distance of $10\,E_R$ from the Earth's center.
(2) the particle touches again the production sphere.
(3) neither 1 or 2 is satisfied before a time limit is reached.

If condition 1 is satisfied the particle is on an allowed trajectory, while if condition 2 is satisfied the particle is on a forbidden one. Condition 3 arises for only a small fraction of the events $O(10^{-6})$.

Particles on allowed trajectories are propagated forward and can reach the Earth's atmosphere. The atmosphere around the Earth is simulated up to 120 Km a.s.l. using 60 concentric layers of homogeneous density and chemical composition. Data on density and chemical composition are taken from the standard MSIS model[11]. The Earth is modeled as a solid sphere which absorbs each particle reaching its surface.

2.1. *The generation technique*

The ideal approach in the generation of the primary cosmic rays spectra would be to start with an isotropic distribution of particles at a great distance (typically $10\,E_R$) from the Earth where the geomagnetic field introduces negligible distortions on

---

[a]The external magnetic field is calculated only for distances greater than 2 Earth's radii (E_R) from the Earth's center . Its contribution to the total magnetic field is $<1\%$ at smaller distances and therefore can be safely neglected.

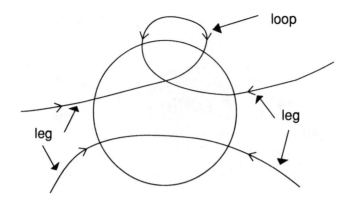

Fig. 1. Trajectories types crossing a spherical surface around the Earth

the interstellar flux. However, this computational method is intrinsically inefficient since most of the particles are generated with trajectories which will not reach the Earth environment. Kinematic cuts can be applied in order to improve the selection efficiency at generation, however they tend to introduce a bias particles with low rigidity.

A good alternative to this approach is the backtracing method[4,14] adopted in the present analysis as outlined in the previous section. In the following, we will shortly discuss the validity of the technique and report the results of a comparison of the two methods. We recall that this method was applied for the first time in ref. [7] for the generation of atmospheric neutrino fluxes.

Let us consider first the effects of the geomagnetic field on an incoming flux of charged particles in the absence of a solid Earth. For the discussion, we start with an isotropic flux of monoenergetic[b] protons at large distance, i.e. at infinity, from the origin of a geocentrical reference system. In this scenario, a negligible fraction of particles, with very particular initial kinematic parameters, will follow complicated paths and remains confined at a given distance from the origin (semi-bounded trajectories); for all practical purposes this sample can be ignored. Most of the particles will follow unbounded trajectories, reaching again infinity after being deflected by the magnetic field.

Unbounded trajectories cross a spherical surface centered in the field source only an even number of times, as shown in Fig.1: we call *legs* the trajectory parts connecting the spherical surface to infinity and *loops* the parts of the trajectory starting and ending inside the spherical surface.

Since each trajectory can be followed in both directions and no source or sink of particles is contained within the surface, the incoming and outgoing fluxes are the same. However, the presence of the magnetic field breaks the isotropy of the

---

[b]The realistic case of an energy spectrum can be treated just as a superposition of monenergetic cases

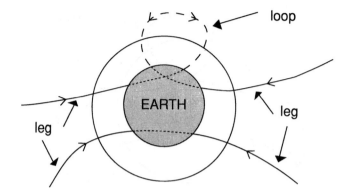

Fig. 2. Trajectories in the presence of a solid Earth

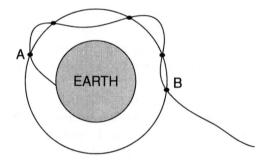

Fig. 3. An example of multiple counting along a trajectory, this type of trajectory has to be considered only at point B.

flux "near" the field source, so for a given location there is a flux dependence due to the direction.

Applying the Liouville Theorem, under the hypothesis of isotropy at infinity, it is straightforward to prove [17] that the proton flux in a random point is the same as at infinity along a set of directions (allowed directions), and zero along all the others (forbidden directions).

The pattern of the allowed and forbidden directions depends on both the rigidity and the location and is known as the geomagnetic cutoff.

With the introduction of a solid Earth, all the trajectories that are crossing the Earth are broken in two or more pieces (Fig.2): the *legs* become one-way trajectories and the *loops* disappear.

The presence of the Earth modifies the flux which exits from the surrounding spherical surface, since particles are absorbed by the Earth, while it has only a minimal effect on the incoming flux which is modified only by the absence of certain *loops*. To generate the flux of particles reaching the Earth's atmosphere, it is sufficient to follow the particles along the allowed trajectories corresponding to the *legs*, taking care to avoid double or multiple counting.

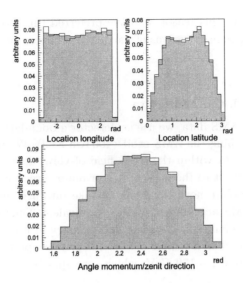

Fig. 4. Latitude and longitude of impact points and angle between momentum and zenith directions for particles generated at a distance of 10 Earth's radii (solid line) and particles generated at 1.07 Earth's radii (shaded histogram).

To respect this prescription we reject all trajectories that are back-traced to the production sphere, this allow us to correctly consider the cases like the one shown in Fig.3.

We point out that an important difference with respect to the application in the neutrino flux calculation of [7] is that for the former, the generation sphere coincided with the Earth's surface, and therefore the forbidden trajectories included those which touched again the Earth (plus those who remained trapped for a long time). In that case there are no problems of double counting.

To check the validity of our technique we made a test comparing the results of the inefficient generation technique at 10 Earth's radii distance from the Earth's center with the backtracing technique described in this paper.

Fig. 4 shows this comparison for several characteristic distributions, the agreement between the two methods is good.

2.2. The interaction model

We use the software package FLUKA 2000 [5] to transport the particles and describe their interactions with Earth's atmosphere.

This package contains a tridimensional description of both electro-magnetic and hadronic interactions. In FLUKA hadronic interactions are treated in a theory-driven approach, and the models and their implementations are guided and checked using experimental data.

This code is benchmarked against a wide set of data and is already used in many applications, ranging from low energy nuclear physics to high energy accelerator and cosmic ray physics. For this reason we have preferred this model with respect to the use of "ad hoc" parametrizations of particle production in the energy range of our interest.

3. Comparison with the AMS data

To compare with the AMS data, we use as detector a boundary corresponding to a spherical surface matching the AMS orbit (400 Km a.s.l). We record each particle that crosses our detector within the AMS field-of-view, defined as a cone with a 32° aperture with respect to the local zenith or nadir directions.

To obtain the absolute normalization, we take into account the field-of-view, the corresponding AMS acceptance, and an Equivalent Time Exposure (E.T.E.) corresponding to the number of the generated primary protons.

Our results are based on a sample of $\sim 6 \cdot 10^6$ primary protons generated in the kinetic energy range of $0.1 - 170\ GeV$, which corresponds to $\sim 4\ 10^{-12}s$ (E.T.E).

3.1. Protons

In Fig.5, we show the comparison between the fluxes obtained with the simulation and the measured AMS downgoing proton flux [2] in nine bins of geomagnetic latitude (θ_M) [12]. Similar results are obtained for the upgoing proton flux.

The simulation well reproduces at all latitudes the high energy part of the spectrum and the falloff in the primary spectrum due to the geomagnetic cutoff, thus validating the general approach used for the generation and detection, as well as the tracing technique.

A good agreement among data and simulation is also found in the under-cutoff part of the spectra. The small and systematic deficit which can be seen in the secondary component of the simulated fluxes is of the same order of the expected contribution from the interaction of cosmic He and heavier nuclei.

This flux is due to the secondaries produced in the atmosphere and that spiral along the geomagnetic field lines up to the detection altitude. Therefore it is sensitive to specific aspects of the interaction model and to the accuracy of the particle transport algorithm.

From the analysis of the motion of the secondary protons from their production up to their detection, it can be pointed out that a fraction of the observed flux is due to a multiple counting of the same particles. Within the formalism of adiabatic invariants [15], it is seen that charged particles trapped in the geomagnetic field, i.e. the undercutoff protons, move along drift shells which can be associated with a characteristic residence time[c] that depends on the fraction of the shell located inside the Earth's atmosphere. Thus, particles moving along *long-lived* shells have a

---

[c]The mean time after which a particle is absorbed into the atmosphere. In our case it represents the effective life time of the particle.

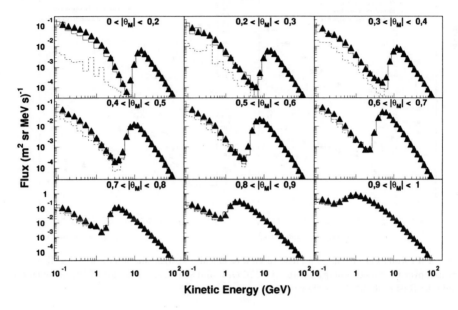

Fig. 5. Downgoing proton flux, simulation(solid line) and the AMS data (triangles); the dashed lines are described in the text. Θ_M is the geomagnetic latitude in radians.

large probability to cross many times a geocentered spherical detector, while those moving along *short-lived* shells typically cross the detector only once.

The drift shells crossing the AMS orbit, at an altitude of 400 km, are in general *short-lived*, however in the equatorial region the *long-lived* shells are present as well [16].

In the following, we will indicate as the *real* proton flux that one obtained by counting only once each particle crossing the detector: its intensity is indicated if fig. 5 by the dashed distributions.

A quite relevant effect can be seen in the equatorial region: there the AMS measurement indicates an important secondary proton flux while the *real* number of protons crossing the detector is more than one order of magnitude lower. At high geomagnetic latitudes, the solid and dashed lines tend to merge. The effect becomes negligible for $\theta_M > 0.8$.

This can be better seen in Fig.6a, where the integral primary proton flux seen by AMS is shown as a function of geomagnetic latitude. The intensities of the *real* and measured undercutoff fluxes are reported in the same plot for comparison and their ratio with the primary component shown in Fig.6b. A minor contribution from the undercutoff proton component can be therefore expected in the atmospheric shower development and neutrino production.

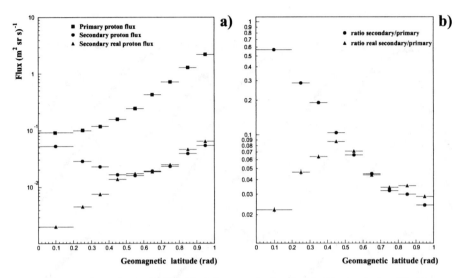

Fig. 6. a) Proton fluxes in the AMS field of view as calculated with this simulation. The fluxes are integrated over the kinetic energy range $0.1-170\ GeV$ and shown as a function of the geomagnetic latitude. b) Ratios of the fluxes shown a).

3.2. Electrons and positrons

In Fig. 7 we show a comparison of the simulated undercutoff electron and positron downgoing fluxes with the corresponding AMS measured fluxes [3].

We remind that the AMS positron measurement is restricted at energies below few GeVs, with a dependence of the maximum energy on the geomagnetic latitude which reflects the increasing proton background with θ_m.

A comparison of data and simulation in the high energy part of the electron spectra is not possible, since the cosmic electrons have not been used as an input in the current work. However, their contribution to the cosmic rays reaching the atmosphere is $O(10^{-2})$ leading to a negligible effect of in the generation of the undercutoff fluxes.

The simulation well reproduces the general behavior of the undercutoff part of the spectra in terms of shape and intensity; a similar agreement is observed for the upgoing lepton spectra (not shown). The *real* lepton fluxes, corresponding to the *real* proton flux described earlier, are shown with the dashed line distribution in Fig. 7. As in the case of protons, a large effect from multiple crossing is present going toward the equatorial region, more pronounced for the positron component.

As for the undercutoff protons, we would have expected a systematic deficit in the simulated electron fluxes coming from the missing contribution of helium and heavier nuclei to the CR fluxes. Subcutoff $e^{\pm}$ are mainly (97%) coming from decays of pions produced in the proton collisions with the atmospheric nuclei: charged pions contribute through the $\pi - \mu - e$ chain, while π^0 through $\pi^0 \to \gamma\gamma$ with subsequent e.m. showers. The relative contribution of charged pions to the subcutoff electrons

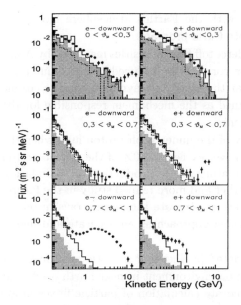
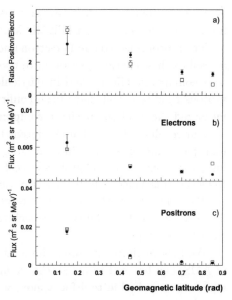

Fig. 7. Downgoing positron and electron fluxes in two regions of geomagnetic latitude θ_M, solid histogram (sim.) black points (AMS data); the hatched histogram and the dashed line distributions are described in the text.

Fig. 8. Electron (b) and positron (c) fluxes and their ratio (a) integrated in the kinetic energy range $0.2 - 1.5\ GeV$, as function of geomagnetic latitude. Open squares (AMS data), black points (simulation).

(positron) fluxes at AMS altitude in our simulation is found to be 37% (47%), while the remaining 60% (50%) appears to come from π^0 production. As suggested in[18] there is the possibility of an overproduction of π^0 but this will be object of more detailed studies.

In Fig.8b-c we show the integrated positron and electron downgoing fluxes for the kinetic energy range $0.2-1.5\ GeV$ as a function of θ_M. Their ratio is shown in Fig.8a. One of the most remarkable features of the AMS measurement is the large value of this ratio, when compared to the natural cosmic value, and its latitude dependence. In Fig. 7, the contribution from primary protons with $E_k > 30\ GeV$ to the electron and positron fluxes is illustrated by the filled area. We can notice that in the equatorial region, the electrons are produced essentially by primary protons with $E_k > 30\ GeV$, while for the positrons lower energy protons contribute as well. This distinction disappears at higher latitudes, where positron and electrons are produced by the protons in the same energy range.

This behavior reflects the East-West asymmetry of the geomagnetic cutoff on primary protons and the larger probabilities of escape from atmosphere for secondary electrons(positrons) generated by Westward(Eastward) moving protons [14,13].

4. Conclusions

The interactions of cosmic ray protons with the Earth's atmosphere have been investigated by means of a fully 3D Monte Carlo program.

The proton, electron and positron undercutoff flux intensities measured by AMS, as well as their energy spectra, have been correctly reproduced by our simulation

Geomagnetic effects, and in particular the east-west asymmetry in the cosmic protons rigidity cutoff, have been confirmed as the mechanism responsible for the measured excess of the positron component.

Our results indicate that the intensity of the undercutoff proton flux, when the multiple counting is taken into account, never exceeds a 10% of the cosmic proton flux, representing a negligible source for atmospheric production of secondaries. However, this aspect will be object of further and more refined study in the future.

The analysis on the possible strategies to generate the cosmic rays incoming flux has shown the validity of a backtracing approach as an accurate and highly efficient technique.

We believe that our simulation, validated by the high statistic measurements of AMS, can be used to assess the radiation environment in near Earth orbit, and represents a valuable tool for a more accurate calculation of particle fluxes in atmosphere.

This work has been partially supported by the Italian Space Agency (ASI) under contract ARS-98/47.

References

1. G. Battistoni et al.,*Astropart. Phys.***12** 315 (2000).
2. J.Alcaraz et al., AMS Collaboration, *Phys. Lett.* **B472** 215 (2000).
3. J.Alcaraz et al., AMS Collaboration, *Phys.Lett.* **B484** 10 (2000).
4. L. Derome, et al. *Phys. Lett.* **B489** 1 (2000).
5. A. Ferrari et al., *Physica Medica* VOL XVII, Suppl. 1.
6. T.K. Gaisser, et al., *Proc. of the 27th ICRC* Session OG1.01 (2001).
7. M. Honda et al., *Phys. Rev.* **D52** 4985 (1995).
8. J.Alcaraz et al., AMS Collaboration, *Phys. Lett.***B490** 27 (2000).
9. N.A. Tsyganenko, *Geomagn. and Aeronomy* **26** 523 (1986);
10. N.A. Tsyganenko and D.P. Stern, *ISTP newsletter,***6** 21 (1996).
11. A. E. Hedin, *J. Geophys. Res.***96** 1151 (1991).
12. G.Gustaffson et al.,*J.Atmos. Terr.Phys***54** 1609 (1992).
13. L.Derome et al., astro-ph/0103474;
14. P.Lipari, astro-ph/0101559;
15. C.E.Mc Ilwain, *J. Geophys. Res.***66** 3681 (1961).
16. E. Fiandrini et al., astro-ph/0106241, E.Fiandrini et al, *Proc. of the XXVII ICRC* (2001).
17. M.S.Vallarta, *Theory of the Geomagnetic Effects of Cosmic Radiation*, published in the Encyclopedia Handbuch der Physik, Springer, Vol. XLVI/1 (1961) 88;
18. G.Battistoni et al., hep-ph/0107241;

REVIEW OF BALLOONS MUON MEASUREMENT IN THE ATMOSPHERE

TOMOYUKI SANUKI

Department of Physics, Graduate school of Science, The University of Tokyo, 7-3-1 Hongo, Bunkyo-ku, Tokyo 113-0033, Japan.

In order to study neutrino oscillation phenomena using atmospheric neutrinos, it is crucially important to calculate their absolute fluxes and spectral shapes accurately. Since production and decay processes of muons are accompanied by neutrino production, observations of atmospheric muons give fundamental information about atmospheric neutrinos. Atmospheric muons have been measured at various sites; from a ground level to a balloon floating altitude. Very precise measurement has been carried out on the ground. Muon growth curves are measured during balloon ascending periods. These data can be used to investigate hadronic interaction models. Investigations of atmospheric muons will improve accuracy of the neutrino calculations. Statistics in the muon measurement during balloon experiments are still insufficient. In order to improve the statistics drastically, dedicated muon experiments are very important.

Keywords: Cosmic Ray; Atmospheric Muon; Atmospheric Neutrino.

1. Introduction

The atmospheric neutrino flux of flavor i (ϕ_{ν_i}) is expressed as

$$\phi_{\nu_i} = \phi_p \otimes R_p \otimes Y_{p \to \nu_i} + \sum_A \phi_A \otimes R_A \otimes Y_{A \to \nu_i}, \qquad (1)$$

where $\phi_{p(A)}$ is the flux of primary protons (nuclei of mass A) outside the influence of the geomagnetic filed. $R_{p(A)}$ and $Y_{p(A) \to \nu_i}$ represent the effect of the geomagnetic field and the yield of neutrinos per primary particle, respectively.[1]. In order to improve accuracy of the atmospheric neutrino calculation, these factors; $\phi_{p(A)}$, $R_{p(A)}$, and $Y_{p(A) \to \nu_i}$ have to be known precisely.

Recently, very precise measurement of primary cosmic-ray spectra below 200 GeV has been carried out by AMS and BESS experiments.[2,3,4] Although the AMS and BESS experiments are fully independent experiments, the resultant spectra show extremely good agreement with each other. The primary cosmic rays in this energy range are relevant to atmospheric neutrinos observed as "fully contained events" in Super-Kamiokande. Thus it seems reasonable to suppose that one already knows the $\phi_{p(A)}$ in Eq. (1) with sufficient accuracy for estimating an event rate of "fully contained events." According to Eq. (1), for accurate calculations of atmospheric neutrinos, it is essential to know the geomagnetic field effect and the neutrino yields besides primary cosmic-ray fluxes.

During space shuttle flight, the AMS experiment directly observed the effect of the geomagnetic field for cosmic rays.[2] On the other hand, it is difficult to observe the effect of the geomagnetic field for secondary cosmic-ray particles such as atmospheric muons. The effect on atmospheric muons would be important for a precise study of low energy atmospheric neutrinos. Cutoff and bending effects in the geomagnetic filed can be deduced from measuring atmospheric muons at different sites. Spectral shapes and charge ratio of atmospheric muons are different depending on the geomagnetic cutoff rigidity.

During ascending period of balloon-borne experiment, a few measurement of atmospheric muons was performed. These experiments show growth curves of muons. Differentiating the muon flux with respect to the atmospheric depth, production cross section and multiplicity of muons could be estimated.

As mentioned above, measuring atmospheric muons provides useful knowledge about the effect of geomagnetic field ($R_{p(A)}$) and the yield of neutrinos per primary particle ($Y_{p(A)\to\nu_i}$). In this article, current status of atmospheric muon measurement is reviewed.

2. Muon Measurement in the Atmosphere

In the atmosphere, muons are being produced and decaying through following processes:

$$p/A + \text{Air} \to \pi + \pi + \cdots, \tag{2}$$

$$\pi \to \mu + \nu_\mu, \tag{3}$$

$$\mu \to e + \nu_e + \nu_\mu. \tag{4}$$

We are observing surviving muons in Eq. (4) on the ground. Climbing up in the atmosphere, muon production processes like Eqs. (2) and (3) become dominant.

Atmospheric muons have been measured at various sites by using ground-based as well as balloon-borne instruments. These flux measurement can be categorized into:

(i) muon measurement on the ground,
(ii) muon measurement during balloon ascending period, and
(iii) muon measurement during balloon floating period.

A thickness of the residual air in each measurement is $1000 - 700$ g/cm^2, $800 - 5$ g/cm^2 and 5 g/cm^2, respectively.

2.1. *On the ground*

2.1.1. *Measurement*

Muons, as well as neutrinos, are the most dominant component among cosmic rays on the ground. In ground-level experiments, there is no limitation of weight

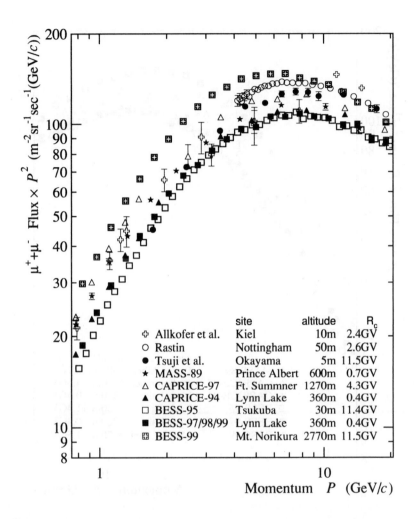

Fig. 1. Atmospheric muon ($\mu^+ + \mu^-$) spectra on the ground.

or power consumption of a detector, which is serious problem for balloon-borne experiments. Therefore, there has been a lot of measurement of atmospheric muons on the ground.[5,6,7,8,9,10,11,12,13,14,15,16].

In Fig. 1, there included relatively new data only. Most of the previous measurement utilized solid iron magnet spectrometers, in which multiple scattering inside the iron made it difficult to measure the absolute rigidity reliably.

An absolute flux is calculated by dividing the number of observed muons in some momentum region by a product of the exposure factor ($S\Omega \cdot t$) and the total

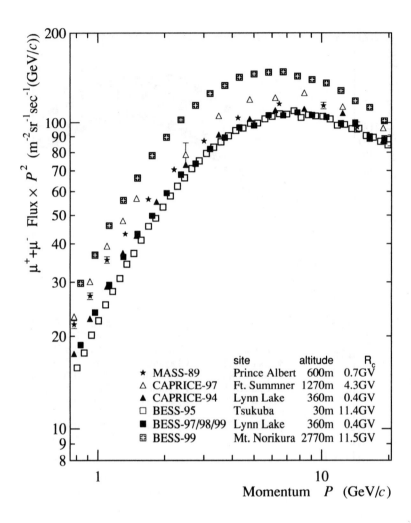

Fig. 2. Absolute fluxes of atmospheric muons ($\mu^+ + \mu^-$) on the ground.

efficiency. Most of the previous experiment did not obtain an absolute flux but normalized their observed spectrum to the "standard" value such as "Rossi point," probably because it was difficult to estimate the exposure factor and the total efficiency precisely.

The "standard" value was usually measured as an integrated flux above some energy with a simple range detector. In these cases, it is not trivial to measure an absolute rigidity of incoming particle event by event. In this kind of normalization, therefore, a small error in momentum measurement leads to a large systematic error

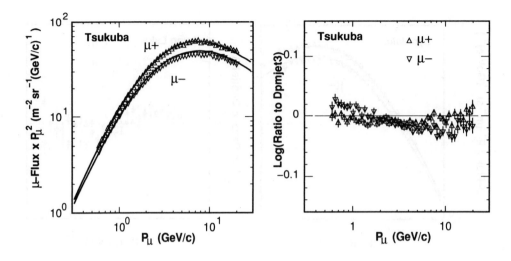

Fig. 3. Observed and calculated muon flux in Tsukuba, Japan (left) and the ratios to the calculation with DPMJET-III model (right).

in the absolute flux.

In Fig. 2, there shown results of absolute flux measurement. In the measurement, superconducting spectrometers were utilized which measured absolute flux without normalization to the "standard" value. A thickness of material inside the superconducting spectrometer is much thiner than the solid iron magnet spectrometers. Systematic errors are therefore well controlled to be small.

The higher an altitude of the experimental site is, the higher flux is observed as shown in Fig. 2. The discrepancy among the absolute fluxes reflects the fact that muon decay process is predominant over production process near sea level.

2.1.2. Calibration of atmospheric neutrino calculation

Errors in resultant spectra measured in ground-level experiments are very small both statistically and systematically. According to recent studies, these precise data are powerful tools to calibrate atmospheric neutrino calculations. Fig. 3 and Fig. 4 demonstrate comparisons between observed and calculated muon spectra in Tsukuba, Japan and Lynn Lake, Canada, respectively.[17] In the calculations, DPMJET-III package [18] was used as a hadronic interaction model. The ratios in Fig. 3 and Fig. 4 show that the fluxes calculated with DPMJET-III agree fairly well with the observed fluxes. The calculation with HKKM interaction model [19] disagree with the observation.[17] Small disagreement between observed and calculated flux is still seen in the ratios. By tuning up the hadronic interaction model, this disagree-

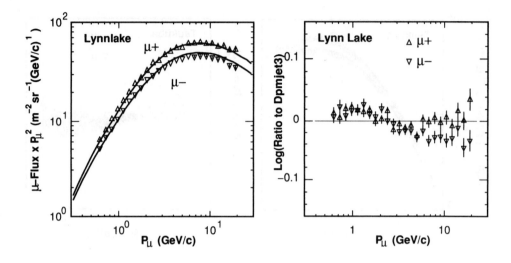

Fig. 4. Observed and calculated muon flux in Lynn Lake, Canada (left) and the ratios to the calculation with DPMJET-III model (right).

ment will disappear. It means that precise measurement of atmospheric muons can calibrate atmospheric neutrino calculation.

2.2. *During balloon ascending period*

When we observe muon fluxes during balloon ascending period, growth curve of atmospheric muons, or correlation between muon intensity and a thickness of the residual air, will be obtained. Since the production and decay process of muons are accompanied by neutrino production as shown in Eqs. (2),(3), and (4), the observation of the muon growth curve is indirect measurement of atmospheric neutrino production. Observing the muon growth curve would be important to investigate hadronic interaction models.

Fig. 5 shows current status of growth curve measurement. From top to bottom are the momentum in GeV/c: 1.0, 1.5, 2.5, 3.9, and 6.3.[16,17,20,21] Overall trends are consistent between observed and calculated growth curves. Statistics in the measurement are too poor to discuss cosmic-ray interactions in the atmosphere. Although ascending speed of the balloon is much slower than space shuttle, it is not slow enough to measure muon growth curves with sufficient statistics. Dedicated muon balloon flights are highly desirable to improve the statistics in muon growth curve measurement.

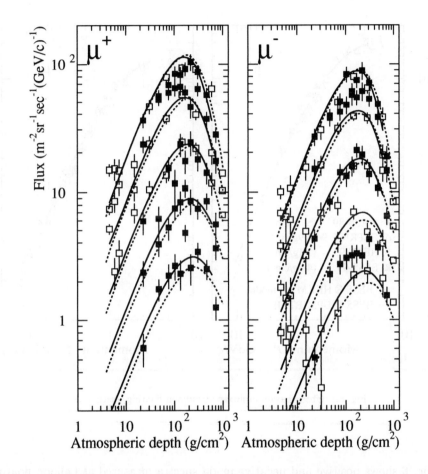

Fig. 5. Atmospheric muon growth curves. From top to bottom are the momentum in GeV/c: 1.0, 1.5, 2.5, 3.9, 6.3. Closed and open squares shows data measured in CAPRICE-1998 and BESS-1998 balloon flights, respectively. Solid and dashed lines are calculated growth curves by R. Engel et al. and M.Honda et al., respectively.

2.3. During balloon floating period

During balloon floating period, a typical thickness of the residual air is 5 g/cm^2, which acts as a "thin" target producing atmospheric muons. Projectile cosmic-ray particle interacts with the target only once. Since muon production process is predominant over decay process at a balloon floating altitude, the measurement is sensitive to hadronic interaction models.

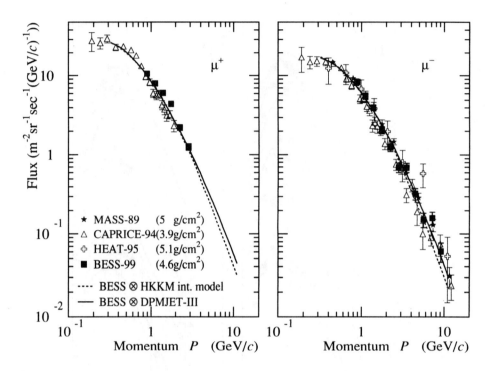

Fig. 6. Observed muon fluxes at balloon floating altitudes.

Fig. 6 shows positive and negative muon spectra measured at balloon floating altitude [16,22,23,24] together with expected muon spectra calculated with HKKM and DPMJET-III interaction models.[17] Because the muon fluxes are very small at the balloon floating altitude, errors in the measurement are still too large to distinguish the two interaction models. Dedicated muon balloon flights are highly desirable to improve the statistics in muon measurement at balloon floating altitude.

3. Future Prospect

Precise measurement of atmospheric muon is indispensable to calibrate or verify the calculations of atmospheric neutrinos. Balloon-borne experiments are unique methods to observe atmospheric muons at any altitudes. Space shuttle or satellite can not observe atmospheric muons.

The muon fluxes have been measured with good statistics on the ground because muons are most dominant component at ground level. By studying these data, hadronic interaction models can be modified.

If we go up to balloon floating altitude, muon flux is as small as only 0.1 % of proton flux. Although long-time measurement by a large instruments with a large aperture is crucially needed, a size of an instrument and flight time is limited in usual balloon flights. Most of the balloon flights had not been designed for muon measurement but for primary cosmic-ray particles.

Dedicated muon balloon flight will be a very important experiment. In such a kind of experiment, it would be important to control balloon ascending speed to be slow enough to measure muon growth curves with sufficient statistics. For lengthening data acquisition time, polar orbit balloon flight is one of ideal balloon experiments.

4. Summary

Atmospheric muons have been measured at various sites; from a ground level to a balloon floating altitude, i.e. the residual air of $1000 - 5$ g/cm^2. Very precise measurement has been carried out on the ground. The measured fluxes have been used to investigate hadronic interaction models. Accuracy of measured absolute muon fluxes in the atmosphere, where a thickness of the residual air is less than 800 g/cm^2, is limited by statistics. In order to improve the statistics drastically, dedicated muon experiment is very important.

Precise measurement of atmospheric muons must be very powerful tools to calibrate the calculations of atmospheric neutrinos. These measurement gives indispensable fundamental data to study neutrino oscillation phenomena using atmospheric neutrinos observed neutrino telescopes such as Super-Kamiokande.

References

1. See, for example, T. K. Gaisser, *Proc. Neutrino Oscillations and their Origin*, ed. Y. Suzuki, M. Nakahara, N.Shiozawa, and K. Kaneyuki (Tokyo, UAP Inc.) ,45 (2000).
2. J. Alcaraz et al., *Phys. Lett.* **B490**, 27 (2000).
3. J. Alcaraz et al., *Phys. Lett.* **B494**, 193 (2000).
4. T. Sanuki et al. *ApJ* **545**, 1135 (2000).
5. P. J. Hayman and A. W. Wolfendale, *Proc. Phys. Soc.* **80**, 710 (1962).
6. B. J. Bateman et al., *Phys. Lett.* **B36**, 144 (1071).
7. O. C. Allkofer, K. Carstensen, and W. D. Dau, *Phys. Lett.* **B36**, 425 (1971).
8. B. C. Nandi and M. S. Sinha, *J. Phys.* **A5**, 1384 (1972).
9. P. J. Green et al., *Phys. Rev.* **D20**, 1598 (1979).
10. A. I. Barbouti and B. C. Rastin, *J. Phys.* **G9**, 1577 (1983).
11. B. C. Rastine, *J. Phys.* **G10**, 1609 (1984).
12. S. Tsuji et al., *J. Phys.* **G24**, 1805 (1998).
13. M. P. De Pascal et al., *J. Geophys. Res.* **98**, 3501 (1993).
14. J. Kremer et al., *Phys. Rev. Lett.* **83**, 4241 (1999).
15. M. Motoki et al., *Proc. 27th ICRC(Hamburg)*, 927 (2001).
16. T. Sanuki et al., *Proc. 27th ICRC(Hamburg)*, 950 (2001).
17. M. Honda et al., *Proc. 27th ICRC(Hamburg)*, 1162 (2001).

18. S. Roseler et al., hep-ph/0012252.
19. M. Honda et al., Phys. Rev. **D54**, 4985 (1995).
20. P. Hansen et al., Proc. 27th ICRC(Hamburg), 921 (2001).
21. R. Engel et al., Proc. 27th ICRC(Hamburg), 1029 (2001).
22. M.T.Brunetti et al., J. Phys **22**, 145 (1996).
23. M. Boezio et al., Phys. Rev. Lett. **82**, 4757 (1999).
24. S. Coutu et al., Proc. ICHEP1998, 666 (1998).

AN ANALYTICAL SOLUTION OF THE COSMIC RAYS TRANSPORT EQUATION IN THE PRESENCE OF THE GEOMAGNETIC FIELD

MARINA GIBILISCO

ECT*, European Centre for Theoretical Studies in Nuclear Physics and Related Areas
Strada delle Tabarelle, 286, 38050 Villazzano, Trento, Italy.

In this work, I study the propagation of cosmic rays inside the magnetic field of the Earth, at distances $d \leq 500\ Km$ from its surface; at these distances, the geomagnetic field deeply influences the diffusion motion of the particles. I compare the different effects of the interplanetary and of the geomagnetic fields, by also discussing their role inside the cosmic rays transport equation; finally, I present an analytical method to solve such an equation through a factorization technique.

Keywords: Cosmic Rays, Propagation

1. Introduction

The propagation of cosmic rays (CR) within the interstellar medium represents a difficult problem, due to the large number of factors potentially influencing the particles motion and connected to their interactions with the electromagnetic fields present in the interstellar plasma. The distribution of cosmic rays within our Galaxy and in its halo is also quite uncertain, the confinement mechanisms due to the galactic magnetic field being not satisfactorily known. It is interesting to investigate the CR propagation in the region extending up to 400 - 500 Km above the Earth surface: in fact, just at these altitudes, the AMS Spectrometer [1] will be operating during its full-regime activity, scheduled for the beginning of 2003. Thus, in view of the future results of the AMS Experiment, the study of particle propagation under the influence of the geomagnetic field seems particularly appealing: I describe here a possible way to analytically solve the CR transport equation in the presence of the geomagnetic field, through a factorization method. A more detailed discussion of this problem can be found in Ref. [2].

2. The Transport Equation for Cosmic Rays

The evolution of the cosmic rays differential number density, $U(\vec{r}, p, t)$ with the time t, the position $\vec{r}$ and the momentum amplitude p of the particle can be described through the following transport equation: [3]

$$\frac{\partial U}{\partial t} + \vec{\nabla} \cdot (\vec{V}U - \mathbf{K} \cdot \vec{\nabla}U) - \frac{1}{3}(\vec{\nabla} \cdot \vec{V})\frac{\partial}{\partial p}(pU) = 0;\qquad(1)$$

In eq. (1), the second term in the left-hand side represents the convection effect: CR particles are convected along the interplanetary magnetic field lines by the solar wind, whose velocity V, in a heliocentric reference frame, reads as follows [4]:

$$V(r) = 400\,[1 - \exp(13.3\,(r_\odot - r))]\;Km\;s^{-1},\qquad(2)$$

and $r = r_\odot = 0.005$ AU is the Sun radius (1 AU $\sim 1.496 \times 10^{11}$ m).

The diffusion tensor $\mathbf{K} = \mathbf{K}(\vec{r}, p, t)$ in the interplanetary magnetic field (IMF) is generally expressed as follows [3,5]:

$$\mathbf{K}(\vec{r},p,t) = \begin{pmatrix} K_\parallel & 0 & 0 \\ 0 & K_\perp & K_T \\ 0 & -K_T & K_\perp \end{pmatrix}.\qquad(3)$$

The coefficients $K_\parallel$, $K_\perp$ and K_T are obtained by assuming that the archimedean spiral pattern for the IMF is dominant, with small fluctuations only [5]: in this case, the standard configuration of the IMF in a heliocentric frame is [3]

$$B_{IMF} = \frac{B_0}{r^2}(\hat{e}_r - \tan\psi\,\hat{e}_\phi),\qquad(4)$$

where ψ is the angle between the direction of the field lines and the radial direction

$$\tan\psi = \Omega\,(r - r_\odot)\frac{\sin\theta}{V};\qquad(5)$$

in eq. (5), $\Omega \sim 3\cdot 10^{-6}\;s^{-1}$ is the angular velocity of the Sun around its axis. Finally, $B_0 = ZA$ is a constant, equal to $\pm 7.5 \times 10^{21}\cdot Z\;gauss\;cm^2$, positive or negative depending on the particular solar cycle we are considering.

At the present time, it is not possible to determine the coefficients $K_\parallel$, $K_\perp$ and K_T with some degree of accuracy and for any level of IMF turbulence [5]; our informations mainly come from the observations, namely from analyses of the intensity and of the time-profiles of solar-CR events [6].

The most suitable form for $K_\parallel$ seems to be [3]:

$$K_\parallel = K_{\parallel 0}\left[1 + \left(\frac{r}{r_e}\right)^2\right]\beta\,\frac{P}{1\;GV},\qquad(6)$$

with $r_e = 1\;AU$, $\beta = v/c$, $0.6 \times 10^{22} \leq K_{\parallel 0} \leq 2.25 \times 10^{22}\;cm^2s^{-1}$ and P is the particle rigidity expressed in GV.

Our knowledge of $K_\perp$ and K_T are equally poor, due to the serious uncertainties concerning their spatial and energy dependence: some possible forms for these coefficients have been obtained within the frame of the quasi-linear diffusion theory (QLT) [3,7].

$$K_\perp = K_{\perp 0}\,\beta\left(\frac{P}{1\;GV}\right)\frac{B_e}{B_{IMF}(r,\theta)},\qquad(7)$$

with $B_e = 5\,nT$;

$$K_T = \pm \frac{\beta P}{3B_{IMF}(r,\theta)}\,f(\theta); \tag{8}$$

(only particles with $Z = 1$ are considered here). In eq. (8) $f(\theta)$ is a transition function modelling the flat neutral sheet, a region where the interplanetary magnetic field is absent; its detailed form and meaning are described in ref.[3].

3. The geomagnetic field

The Earth magnetosphere has a very complex structure, which can be fundamentally divided into two regions, although presenting some irregularities; we distinguish:

a) the main geomagnetic field, decreasing in a very sensitive way with the distance from the Earth surface and approximately represented by a dipole field, tilted by an angle of 11.4° with respect to the Earth rotation axis; small diurnal and seasonal effects, producing irregularities, are caused by its interaction with the solar wind;

b) the magnetopause, namely the surface boundary of the geomagnetic field, along which electric currents are flowing, having the net effect to suppress the field outside this boundary. The general configuration of the magnetopause is not simply evaluable, being determined by a complex balance between the solar wind pressure and the field pressure.

Some irregularities are present in this configuration, thus the description of the geomagnetic field generally involves a large number of parameters[8].

Finally, three calculation strategies can be adopted, depending on the particular description of the magnetopause one assumes:

1) the magnetosphere configuration can be obtained by directly calculating the magnetospheric boundary;

2) the same configuration can be obtained by $a - priori$ choosing a particular shape for the magnetopause;

3) finally, one can perform a numerical least-squares fit of the *measured* magnetic field.

In my calculation, I will adopt the last method: the resulting geomagnetic field model involves a multipole expansion of the spherical harmonic function[9]:

$$B_{north} = \frac{1}{r}\frac{\partial V}{\partial \theta} \qquad B_{east} = -\frac{1}{\sin\theta}\frac{\partial V}{\partial \phi} \qquad B_{down} = \frac{\partial V}{\partial r}; \tag{9}$$

the potential function reads [9]

$$V = R\sum_{n=1}^{\bar{n}}\sum_{m=0}^{n}\left(\frac{R}{r}\right)^{n+1}\left(g_n^m\cos(m\phi) + h_n^m\sin(m\phi)\right)P_n^m(\cos\theta), \tag{10}$$

where R is the Earth radius and $P_n^m(x)$ is the associated Legendre function. The coefficients of the expansion, (g_n^m, h_n^m) are listed in ref.[10].

The field expressed by eq. (9) and (10) has been originally evaluated by a FLUKA Routine (courtesy of C. Bloise and G. Battistoni), in a cartesian reference frame, having its origin at the Earth centre: the y axis points towards the geographic North Pole, the z axis is directed towards the Greenwich meridian and, finally, the x axis completes the triad.

I modified the FLUKA routine by adopting a spherical coordinate frame and by calculating the partial derivatives of the local field: if we know the local values of the geomagnetic field, the transport equation, in the presence of such a field, is similar to eq. (1) but it should be written in a geocentric frame: that can be done by performing a simple translation by a distance equal to the Sun-Earth separation.

In eq. (1), the convection term is not modified by the presence of the geomagnetic field: in fact, convection is strictly connected to the solar wind, fundamentally an electron/proton gas which traps and freezes out the interstellar magnetic field, along whose lines the CR particles are convected. The relevant quantity governing this motion is therefore the solar wind velocity, eq. (2), also appearing in the last term of eq. (1), where it influences the energy change of the particles in fluxes for which $\vec{\nabla} \cdot \vec{V}$ is not vanishing.

The most relevant modification of the transport equation in the presence of the geomagnetic field concerns the diffusive motion; in proximity of the Earth (namely, at distances $d \leq 500\ Km$ from its surface), B_{GEO} is the dominant field, thus the problem of the diffusion should be reconsidered: in fact, the parametrization for the diffusion tensor expressed by the matrix (3) is obtained by taking into account the IMF field only, in the specific case of an archimedean pattern for the IMF field, with fluctuations relatively small [5].

Here, I will consider the diffusion problem from a local point of view: by neglecting the possible CR interactions with the Earth atmosphere and the possible ionization processes, the diffusion can be locally described through a single coefficient $K(r, \theta, \phi, p)$ which, however, maintains the observed characteristic behaviour with respect to the charge, to the velocity and to the momentum (or to the rigidity) of the particles.

The most suitable form for it is:

$$K(r,\theta,\phi,p) = \frac{K_0}{Z} \left(\frac{p^2}{m\ GeV}\right) \frac{B_{IMF}(Earth\ Surf.)}{B_{GEO}(r,\theta,\phi)}, \qquad (11)$$

where the constant $K_0 = 2.2 \cdot 10^{24}\ cm^2/sec$ comes from the experimental observations [3].

Eq. (11) is expressed as a function of the particle momentum p instead of the rigidity R: in such a way, it is possible to explicitly introduce, in the transport equation, the particle mass and the charge dependence, so potentially discriminating matter from antimatter. In eq. (11), B_{IMF}, evaluated at the Earth surface, acts as a weight, measuring the relative influence of the interplanetary field with respect to the geomagnetic one, as in $K_\perp$ (see eq. (7)). More involved interference effects between the two magnetic fields could be considered in future.

Finally, if we select CR particles having an assigned initial momentum p_0, it is possible to skip the p-dependence in eq. (11), thus further simplifying the problem.

4. An analytical solution of the cosmic rays transport equation

By explicitly writing all the spatial derivatives, the cosmic rays transport equation can be put in a compact form; if we define the following functions as

$$A(r) = \left(\frac{2 V_r}{r} + \frac{\partial V_r}{\partial r} \right), \tag{12}$$

$$B(r,\theta,\phi) = V_r - \frac{\partial K}{\partial r}, \tag{13}$$

$$C(r,\theta,\phi) = \left(\frac{1}{r^2} \frac{\partial K}{\partial \theta} + \frac{K}{r^2} \cot\theta \right), \tag{14}$$

$$D(r,\theta,\phi) = \left(\frac{1}{r^2 \sin^2\theta} \frac{\partial K}{\partial \phi} \right), \tag{15}$$

the transport equation reads:

$$\frac{\partial U}{\partial t} + \frac{2}{3} A(r) U - \frac{1}{3} A(r) p \frac{\partial U}{\partial p} + B(r,\theta,\phi) \frac{\partial U}{\partial r} - C(r,\theta,\phi) \frac{\partial U}{\partial \theta} - D(r,\theta,\phi) \frac{\partial U}{\partial \phi} -$$
$$- K(r,\theta,\phi) \cdot \left[\frac{\partial^2 U}{\partial r^2} + \frac{1}{r^2} \frac{\partial^2 U}{\partial \theta^2} + \frac{1}{r^2 \sin^2\theta} \frac{\partial^2 U}{\partial \phi^2} \right] = 0, \tag{16}$$

where V is the solar wind velocity (written in a geocentric frame) and K is the diffusion coefficient, given by eq. (11); here, I consider, as an example, particles having an initial momentum $p_0 = 1.2\ GeV$.

I search for a factorized solution of eq. (16), having the following form:

$$U(r,\theta,\phi,p,t) = F(t)\,G(p)\,H(r,\theta,\phi); \tag{17}$$

By substituting the right-hand side of eq.(17) in the original equation and by dividing all the terms by $F(t)G(p)H(r,\theta,\phi)$ it becomes possible to separate the dependence of the solution on the time and on the momentum from the dependence on the spatial variables. Then, the final equation reads:

$$\frac{1}{F} \frac{\partial F}{\partial t} + \frac{2}{3} A(r) - \frac{1}{3} p\, A(r) \frac{1}{G} \frac{\partial G}{\partial p} + \frac{1}{H}\left[B(r,\theta,\phi) \frac{\partial H}{\partial r} - C(r,\theta,\phi) \frac{\partial H}{\partial \theta} -\right.$$
$$\left. - D(r,\theta,\phi) \frac{\partial H}{\partial \phi} - K(r,\theta,\phi) \left[\frac{\partial^2 H}{\partial r^2} + \frac{1}{r^2} \frac{\partial^2 H}{\partial \theta^2} + \frac{1}{r^2 \sin^2\theta} \frac{\partial^2 H}{\partial \phi^2} \right] \right] = 0. \tag{18}$$

The time-dependence of the solution also separates as follows:

$$\frac{1}{F(t)} \frac{\partial F}{\partial t} = a_1, \tag{19}$$

$$F(t) = exp(a_1\ t) + a_2, \qquad (20)$$

with $a_1 < 0$, in order to avoid divergences.

Multiplying all the terms by $\frac{3}{A(r)}$, the momentum term also factorizes:

$$2 - p\,\frac{1}{G(p)}\,\frac{\partial G}{\partial p} = b_1, \qquad (21)$$

$$G(p) = b_2\,p^{(2-b1)}. \qquad (22)$$

Eq. (22) tells us that $G(p)$ has a power-law dependence on the momentum, as one expects on the basis of the experimentally observed spectra.

Again, the measurements of the fluxes and of the particle spectra enable us to fix the boundary conditions and to evaluate the integration constants a_1, a_2, b_1, b_2.

For instance, the estimated proton flux at $E_{kin} = 1\ GeV$ is [11]

$$F_p = 0.8\ m^{-2}\ sr^{-1}\ sec^{-1}\ MeV^{-1}.$$

This value, together with the observed energy dependence of the flux (expressed by an inverse power law with an index equal to -2.7), suggest the following choice for the integration constants:

$$a_1 = -1.93 \times 10^{-10}\ sec^{-1}, \qquad (23)$$

$$a_2 = 0, \qquad (24)$$

$$b_1 = 4.7, \qquad (25)$$

$$b_2 = 5.58 \times 10^4. \qquad (26)$$

The value (25) comes from the observed power-law energy dependence; (26) is obtained from eq. (22) with the particle flux estimated at $1\ GeV$; (24) is obtained by imposing that, for $t \to \infty$ and $a_1 < 0$, the flux vanishes; finally, (23) is calculated from eq. (20), which is estimated at a time $t \sim 1.16 \times 10^9\ sec$, corresponding to the ideal time during which the particles cross our Galaxy from its boundaries up to the solar system.

The remaining equation, having as a variable the function $H(r, \theta, \phi)$, can be solved by supposing again that our solution factorizes into the following product of two functions, respectively depending on the radial distance r and on the angular variables (θ, ϕ):

$$H(r, \theta, \phi) = S(r)\,Y(\theta, \phi); \qquad (27)$$

in this way, I obtain:

$$-b_1 + \frac{3}{A(r)}\left[-a_1 - \frac{B(r,\theta,\phi)}{S(r)}\frac{\partial S}{\partial r} + \frac{C(r,\theta,\phi)}{Y(\theta,\phi)}\frac{\partial Y}{\partial \theta} + \frac{D(r,\theta,\phi)}{Y(\theta,\phi)}\frac{\partial Y}{\partial \phi} + \right.$$

$$\left. + K(r,\theta,\phi)\left[\frac{1}{S}\frac{\partial^2 S}{\partial r^2} + \frac{1}{r^2}\frac{1}{Y(\theta,\phi)}\frac{\partial^2 Y}{\partial \theta^2} + \frac{1}{r^2 \sin^2\theta}\frac{1}{Y(\theta,\phi)}\frac{\partial^2 Y}{\partial \phi^2}\right]\right] = 0. (28)$$

We may assume that the angular dependence of the solution traces the geomagnetic field behaviour:

$$Y_n^m = \sin(m\phi) P_n^m(\cos\theta), \qquad (29)$$

Then, calling:

$$Z(\theta,\phi) = \frac{1}{Y}\frac{\partial Y}{\partial \theta}, \qquad (30)$$

$$W(\theta,\phi) = \frac{1}{Y}\frac{\partial Y}{\partial \phi}, \qquad (31)$$

$$M(\theta,\phi) = \frac{1}{Y}\frac{\partial^2 Y}{\partial \theta^2}, \qquad (32)$$

$$N(\theta,\phi) = \frac{1}{Y}\frac{\partial^2 Y}{\partial \phi^2}, \qquad (33)$$

and

$$P(r,\theta,\phi) = -b_1 - \frac{3a_1}{A(r)} + \frac{3}{A(r)}\left[C(r,\theta,\phi)Z(\theta,\phi) + D(r,\theta,\phi)W(\theta,\phi)\right] +$$
$$+ \frac{3}{A(r)} K(r,\theta,\phi)\left[\frac{1}{r^2} M(\theta,\phi) + \frac{1}{r^2 \sin^2\theta} N(\theta,\phi)\right], \qquad (34)$$

the transport equation reads:

$$P(r,\theta,\phi) - \frac{3}{A(r)}\left[\frac{B(r,\theta,\phi)}{S}\frac{\partial S}{\partial r} - K(r,\theta,\phi)\frac{1}{S}\frac{\partial^2 S}{\partial r^2}\right] = 0. \qquad (35)$$

Finally, by calling:

$$\frac{1}{S}\frac{\partial S}{\partial r} = T(r), \qquad (36)$$

eq. (28) transforms into a more simple, Riccati-like differential equation, which can be simply solved both in an analytical and in a numerical way (for instance through Runge-Kutta-Merson methods).

Better approximations of the general solution may be obtained by repeating such a procedure in many recursive steps, by considering the $(n-1)$-order solution as a particular solution of the Riccati-like equation at the step n: the work is in progress to test such a possibility, which would make the factorization approach more soft.

The solution I obtained through the factorization method previously described are shown in fig. 1 (where I plot the function $T(r)$) and in fig. 2 ($S(r)$ is shown, as obtained by solving eq. (36)).

The most interesting application of these studies consists in the possibility to find some traces of cosmological antimatter in cosmic rays: in fact, particles having different charges and masses are differently deviated by a magnetic field and, therefore, both accurate flux measurements and theoretical estimations of the particle differential number density may be useful in this sense.

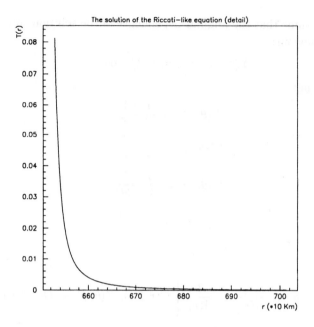

Fig. 1. The solution of the Riccati-like equation, $T(r)$.

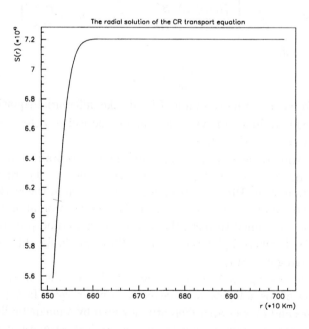

Fig. 2. The radial solution of the cosmic rays transfer equation, $S(r)$.

References

1. Ahlen, S. P. et al., *Nucl. Instrum. Methods* **A350**, 351, (1994).
2. M. Gibilisco, *International Journal of Mod. Phys.* **A16**, 2293, (2001).
3. Potgieter,M. S., Moraal, H., *ApJ* **294**, 425, (1985).
4. Collard, H. P., Milhalov, J. D., Wolfe, J. H., *Journ. of Geophys. Res.* **87**, 2203, (1982).
5. Quenby, J. J., *Space Science Rev.* **37**, 201, (1984).
6. Palmer, I. D., *Rev. Geophys. Space Phys.* **20**, 335, (1982).
7. Forman, M. A., Jokipii, J. R., Owens, A. J., *ApJ* **192**, 535, (1974).
8. Kosik, J. C., *J. Geophs. Res.* **94**, 12021, (1989); and *Ann. Geophys.* **16**, 1557, (1998).
9. Honda, M., Kajita, T., Kasahara, K., Midorikawa, S., *Phys. Rev.* **D52**, 4985, (1995).
10. Barraclough, D. R. et al., *J. Geomagn. Geoelectr.* **37**, 1157, (1985).
11. Ip, W. H., Axford, W. I., *Astron. Astrophys.* **149**, 7, (1985).

Interaction of Cosmic Rays with the Geomagnetic Field

Chairperson: G. Battistoni

LEPTONS WITH E>200 MEV TRAPPED IN THE SOUTH ATLANTIC ANOMALY

E. FIANDRINI

Università di Perugia e Sezione INFN
Via Pascoli, 06123 Perugia, Italy

G. ESPOSITO

Università di Perugia e Sezione INFN
Via Pascoli, 06123 Perugia, Italy

For the first time accurate measurements of electron and positron fluxes in the energy range 0.2÷10 GeV have been performed with the Alpha Magnetic Spectrometer (AMS) at altitudes of 370÷390 km in the geographic latitude interval ±51.7°. We focused on the under-cutoff lepton fluxes inside the region of the South Atlantic Anomaly (SAA), defined as region where the local magnetic field B $\leq$ 0.26 G.

A clear transition region from stably-trapped flux typical of the Inner Van Allen Belts to quasi-trapped flux typical underneath the Van Allen Belts is observed in the SAA up to energies O(Gev).

The observations strongly support positrons abundance in the Inner Van Allen Belts, both in the stably trapped component and in the quasi-trapped one.

The flux maps as a function of the canonical adiabatic variables L, α_o are presented for the interval 0.95 < L < 3, 0° < α_o < 90° for electrons (E<10 GeV) and positrons (E<3 GeV). The results are compared with existing data at lower energies.

Keywords: leptons, radiation belts, AMS

1. Introduction

The presence of high energy (up to few hundred MeV) electrons and positrons trapped in the Inner Van Allen Belts is well known. The existing experimental data in the energy range of 0.04÷200 MeV come from satellites covering a large range of adiabatic variables [15,4,1,8].

Although the magnetic trapping mechanism is well understood, a complete description of the phenomena, including the mechanisms responsible for the injection and depletion of the belts as well as those determining the energy spectra is lacking. This is particularly true for the the region of the SAA where the Inner Belts come closer to Earth's atmosphere and may interact with it.

At lower energies, models are available for leptons (up to few MeV) and protons (up to hundreds MeV) [14,6] based on the data provided by satellite campaigns, continuosly updated for istance in the context of the Trapped Radiation ENvironment Development project[16].

At higher energies the existing data come from measurements carried out by the Moscow Engineering Physics Institute. These data, taken at altitudes ranging from 300÷1000 km with different instruments placed on satellites and the Mir station [15,4,1], established the existence of O(100 MeV) trapped leptons both in the Inner Van Allen Belts (*stably* trapped) and in the region below (*quasi-trapped*), and determined their charge composition[3,5]. At these altitudes, the shell structure is strongly distorted in the vicinity of the SAA, and consequently the observations are sensitive to different regions of trapped particles: the Inner Van Allen belts over the SAA and quasi-trapping belts outside of the SAA.

The Russian measurements concern mainly the region of the SAA; very little data is available at the corresponding altitudes outside the SAA.

The measured ratio of e^+ to e^- is found to depend strongly on the observed population type. In the SAA, electrons dominate the positrons by a factor $\sim$ 10, a ratio similar to that observed for the cosmic fluxes, while outside the SAA the two fluxes are similar and comparable to the e^+ flux inside the SAA[5]. However, the situation is not completely clear, since other groups report a lower e^- excess ($\sim$ 2) for the SAA[10].

In this paper, we use the high statistics data sample collected by the AMS experiment in 1998, for a detailed study of the high energy (O(1 GeV)) lepton fluxes in the South Atlantic Anomaly (SAA) region. The SAA is defined by the value of the local magnetic field, as the region where B$\leq$ 0.26 G, corresponding to the geographic area shown in fig 2.

The data are analyzed in terms of the canonical invariant coordinates characterizing the particle motion in the magnetic field: the L shell parameter, the equivalent magnetic equatorial radius of the shell, the equatorial pitch angle, α_0, of the momentum $\vec{p}$ with the $\vec{B}$ field, and the mirror field B_m at which the motion reflection occurs during bouncing[12,9].

2. AMS Data Analysis

The Alpha Magnetic Spectrometer (AMS) is equipped with a double-sided silicon microstrip tracker, with an analyzing power of BD^2=0.14 Tm^2, being B the magnetic field intensity and D the typical path length in the field. Details on the detector performance, lepton selection and background estimation can be found in [2] and references therein.

The AMS was operated on the shuttle Discovery during a 10-day flight, beginning on June 2, 1998 (NASA mission STS-91). The orbital inclination was 51.7° in geographic coordinates, at a geodesic altitude of 370÷390 Km. The detector, not magnetically stabilized, recorded data for $\sim$20 hours in the region of SAA and for $\sim$ 44 hours outside of it with different fixed attitudes with respect to the local zenith direction (0°, 20°, 45°, and 180°). The corresponding orbital coverage in geographical coordinates is shown in Fig. 2 for SAA.

The values of L, α_0 and B_m of the detected leptons were calculated using the

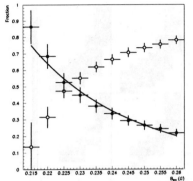 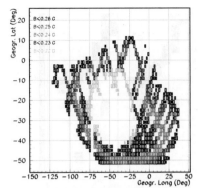

Fig. 1. Fraction of *stably-trapped* (full circles) and *quasi-trapped* (empty circles) as a function of the maximum local magnetic field B_{SAA}, used to define SAA contours ($B \leq B_{SAA}$). The power law used for the fit is superimposed.

Fig. 2. Orbital coverage corresponding to the geographical area where the local magnetic field B is less than 0.26 gauss. Different colors indicate different values of the max local magnetic field.

UNILIB package [16] with a realistic magnetic field model, including both the internal and the external contributions[11,13]. More details can be found in [7]. The AMS Field of View (FoV) along the orbits in the SAA is shown in Fig.6.

Trigger rates varied between 100 and 700 Hz, with maximum rate in the core of SAA (B≤ 0.205 G) where detector livetime was very low and AMS did not take data. Data analyzed here refer therefore to the edges of the SAA region where measurements are still reliable, with a trigger efficiency ≥ 0.9. The region of count rate saturation can be seen in fig. 2 as the empty region in the inner region of SAA.

To reject the cosmic component of the measured lepton fluxes, the lepton trajectories in the Earth's magnetic field were traced taking realistic geomagnetic field models and the Earth's penumbra effects into account, as described in more detail in [7].

Particles are classified as cosmic if they reach infinity from Earth, secondaries if they don't; if classified secondary, the particle may reach the Earth's atmosphere at some point of its trajectory or it can stay indefinitely in flight around the Earth, i.e. for more than a fixed time, taken as 30 s in our case: at 90 of CL if a particle traveled for 30 s, it will travel indefinitely. These two components are said *Quasi-Trapped* (QT) and *Stably-Trapped* (ST), respectively.

The fraction of a given type of component is a strong function of the maximum field value used to define the SAA region, $B \leq B_{SAA}$, as shown in fig. 1. The fraction of stably trapped component, F_{st}, can be fitted with a power law of B_{SAA}, as $F_{st} = (B_{SAA}/B_o)^{-\alpha}$, with Bo=0.19 G, α=6.9 and the quasi trapped as $F_{qt} = 1 - F_{st}$. B_o is the field value at which the flux becomes entirely stably trapped (F_{st}=1) and represents the limit of the full Inner Van Allen (IVA) belts at the altitude of AMS orbit. Because of low detector livetime in the central region of SAA, the lowest observable field value is B~0.21 G and the observed stably trapped component

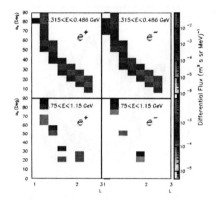 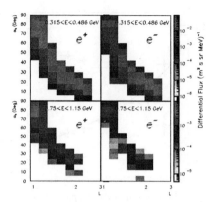

Fig. 3. e^+ and e^- differential flux maps for the *Stably-Trapped* (ST) component inside the SAA, for two different energy bins.

Fig. 4. e^+ and e^- differential flux maps for the *Quasi-Trapped* (ST) component inside the SAA, for two different energy bins.

fraction is less than 1. Therefore, we can not observe the full closed IVA belts, but rather the transition region between IVA and the regions underneath them, where drift shells are not closed, a sort of 'gray' spot between two different geomagnetic regions, which we called *Transition Radiation Belts* (TRB).

Because of the observed mixing between QT and ST components, this TR belt region could responsible for the filling and depletion of IVA, especially at high energies, due to possible pitch angle scattering by elastic interactions with residual atmosphere before or after detection in AMS.

3. AMS Results

For the description of fluxes, the canonical total energy E, L-shell parameter and the equatorial pitch angle α_0 were used (this is preferred to B_m since limited to $0° \div 90°$). A three-dimensional grid (E, L, α_0) was defined to build flux maps.

Because of the observed two-fold nature of the particles in the TRB, it is worth to study separately the two components. Two differential flux maps in (L, α_0) at constant E are given in Figs. 3 and 4, for the QT and ST leptonic components inside the SAA and in Fig. 5 for the QT flux outside the SAA, respectively.

These distributions show that the QT component fills all the available phase space inside the SAA both in the low and high energy part of the differential spectrum, while a much narrower distribution is observed for the ST component, which lies all in a narrow band corresponding to inner regions of the SAA.

The lower limit for the stably trapped component can be fitted with the relation $sin\alpha_{st} = 0.95/L^2$. This is the limit for stable trapping in the Inner Van Allen belts at the altitude of AMS, in the sense particles with $\alpha_o \leq \alpha_{st}$ *cannot* be injected into the Inner Van Allen belts; the angle α_{st} can be defined as the *equatorial drift loss cone angle*.

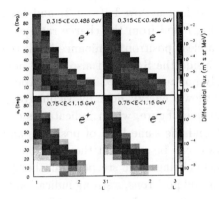

Fig. 5. e^+ and e^- differential flux maps for the *Quasi-Trapped* (ST) component outside the SAA, for two different energy bins.

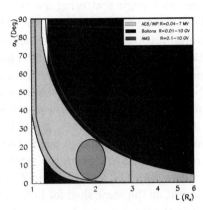

Fig. 6. Comparison of the field of view of AMS inside (contour) and outside (gray region) SAA with balloons and satellite measurements in (L, α_o).

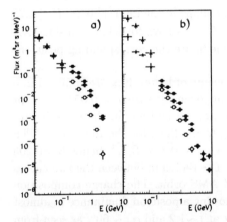

Fig. 7. Differential flux of positrons (a) and electrons (b) in the region $L \leq 1.2$ and $\alpha_o \geq 70°$. Triangles and stars show Mariya data inside and outside the SAA, respectively. Full and empty circles show AMS quasi-trapped and stably-trapped component inside the SAA, respectively. The crosses show the quasi-trapped component outside the SAA, measured by AMS.

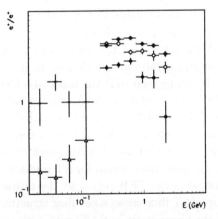

Fig. 8. Charge ratio of positrons to electrons in the region $L \leq 1.2$ and $\alpha_o \geq 70°$. Triangles and stars show the ratio for Mariya data inside and outside the SAA, respectively; the empty and full circles show the measured AMS quasi-trapped and stably-trapped ratio, respectively; the crosses show the measured AMS ratio outside the SAA.

The ST component is concentrated at low energies, being the flux at high energy very low and sparse in the available phase space. This is also a clear indication that the trapping mechanism works only at relatively low energies. Essentially for particles with E above some hundreds of MeV, the trapping becomes very unlikely because pitch angle scattering due to Coulomb interactions is less important.

This can be seen also from the differential flux at low L ($\leq$ 1.2) and high equatorial pitch angles ($\alpha \geq 70^o$) shown in Fig.7 and from the charge ratio as a function of energy, shown in Fig.8. There is a clear positron dominance in all the components observed by AMS, ST and QT inside the SAA and also in the not-stably component outside the SAA, as already pointed out in [7]. The observations are in contrast with previous measurements at lower energy, done by Mephi group with the Mariya instrument[4], plotted on the same figure, which indicate instead a strong electron abundance inside the SAA and the same level of positrons and electrons outside SAA. From the same figures it can be seen that the flux slope is very different for QT and ST: the ST components being much steeper than the QT component on the same shell for both electrons and positrons, a strong indication of a limit in the trapping mechanism in Van Allen belts. Another important feature is that the differential QT fluxes inside and outside the SAA have the same structure, slope and comparable intensity: a clear indication that these two components are essentially the same population of particle observed on different points of the same shell, as can be seen from Figs. 4 and 5.

The integral flux is useful to have a picture of the overall distribution of the particles in the belts. In Fig. 9 the integral flux maps in the interval $0.2 \leq E \leq 2.73$ GeV for electrons (a, b, c) and positrons (d, e, f) are shown for ST, QT inside SAA and QT outside SAA, respectively. In the same figure the corresponding positrons to electrons ratio are plotted (g, h, i).

It can be seen that the integral ST components of lepton flux are confined in a thin band around the SAA, with a lower limit given by $sin\alpha_{st} = 0.95/L^2$, as for the differential flux and shown as full line in figs. 9a and 9d. In the same figures is also shown the dashed curve corresponding to the limit for the Inner Van Allen belts at the altitude of AMS, given by $sin\alpha_{IVA} = \sqrt{0.311/B_oL^3}$, being B_o=0.205 G, the same value found for ST fraction, F_{st}. The region in between the two curves represents the TRB region at the altitude of AMS orbit. The charge composition map of ST flux shows a peculiar structure, showing a positron dominance in almost all the phase space in SAA, with a clear peak at L~1.2 and $\alpha_o \sim 60^o$, as seen from fig. 9g.

The QT component of the flux looks quite different from the ST one: it fills all the phase space in SAA and not confined in a narrow band as the ST flux, as shown in figs. 9b and 9e. The structure of QT flux inside and outside SAA looks very similar in structure and intensity, indicating the common nature for the two fluxes: the same particle population observed in different locations of the drift shell, as already mentioned. This can be seen more clearly also from the charge ratio maps in figs. 9h and 9i, where is clear that for both inside and outside there is a maximum at low L and high α_o.

Integration over α_o gives the flux as function of L shell parameter and describes the radial distribution of flux in the belts, as shown in fig. 10. From the figure 10b and 10c, it can be seen that the radial distribution of the flux inside and outside

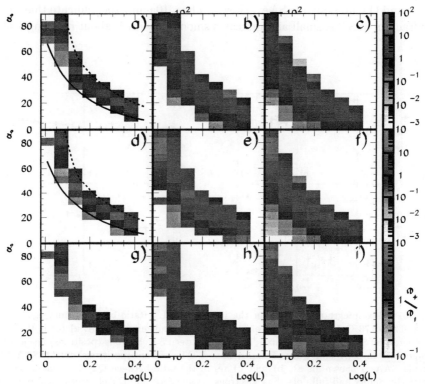

Fig. 9. Integral flux in the interval 0.205-2.73 GeV for Stably Trapped (a), Quasi Trapped (b) in SAA and Not Stably Trapped outside the SAA (c) electrons; for Stably Trapped (d), Quasi Trapped (e) in SAA and Not Stably Trapped outside the SAA (f) positrons; the corresponding charge ratios are shown in g), h) and i), respectively. The dotted curve shows the limit for the full Inner Van Allen belts, the full line shows the limit for stable trapping at AMS orbit altitude, as explained in text.

the SAA is very similar, as the charge ratio (e, f) for electrons (empty dots) and positrons (full dots), again inferring the common nature of the two fluxes. This is clearly seen in fig. 11b and 11d, where the ratio between same sign leptons inside and outside SAA is plotted as a function of L: it depends weakly on L and ratio smoothly increases from unity at low L to about 2 at high L.

It must be remarked that the geographical area corresponding to the region of the SAA is very limited compared to the area outside, as shown in fig. 2, but the L coverage is about the same, because on the edges of SAA the drift shells are dropping down to very low altitudes due to the local magnetic field deformation and span a large L range in a limited geographical region.

Differently, the radial distribution of the ST flux shows a clear structure for both electrons and positrons, as shown in fig. 10a with two maxima at $L \sim 1.2$ and

~ 2.1 and a minimum in between at L ~ 1.4; the ratio positrons to electrons, shown in fig. 10d, has a maximum in the region of minimum flux. The ratio between ST inside and QT flux outside SAA , shown in figs. 10a and 10c, indicate that at low L the fraction of ST is small and increases smoothly with L about reaching 1 at high L.

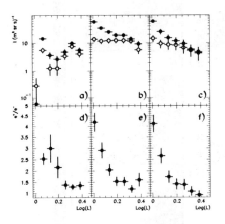

Fig. 10. Flux integrated over α_0 in the interval 0.205-2.73 GeV as a function of L shell parameter for Stably trapped, Quasi Trapped components in SAA and Quasi Trapped flux outside SAA is shown in a), b) and c) respectively; empty and full dots show electrons and positrons, respectively; the corresponding charge ratios are shown in d), f) and e) respectively.

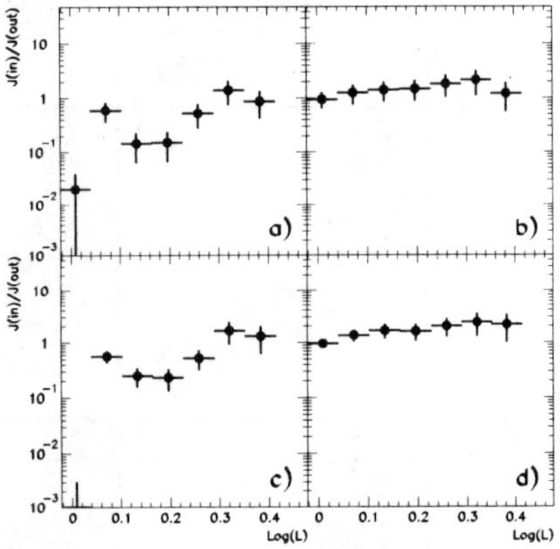

Fig. 11. Ratio between same sign leptons inside and outside SAA. a) (c) is the ratio between ST electrons (positrons) inside SAA and QT electrons (positrons) outside SAA. b) (d) is the ratio between QT electrons (positrons) inside SAA and QT electrons (positrons) outside SAA.

4. Discussion

For the first time the region between the full Inner Van Allen belts and underneath them have been measured by AMS experiment up the energies of several GeVs.

The observations support the existence of a transition region from stably trapped to quasi trapped populations at the altitude of AMS orbit. The transition region is defined by the curve marking the lower limit for stable trapping, given by the equatorial drift loss cone angle and the curve for the upper limit of the full Inner Van Allen belts. This region, called here Transition Radiation Belt, could be responsible for the injection and loss of particle into the Inner Van Allen belts, due the mixing of ST flux, typical of Inner Van Allen belts and QT flux, typical of the regions beneath the belts. The mixing is likely due to Coulomb scattering with residual atmosphere before and after detection, which partially randomizes particle pitch angles and make them to drift in closed shells. The amount of mixing depends on the average amount of matter transversed by the particle during its motion, which turns out to be of the order of 5-10 X_o of air.

The features of the two components are different: ST flux shows a peculiar structure with two maxima and a slot region in between and limited to relatively low energy ($\sim$ hundreds of MeV) by the geomagnetic cutoff; it is detected only in a narrow region corresponding to the inner regions of SAA observed by AMS.

The QT inside SAA component seems to be very similar to QT flux outside the SAA, strongly supporting the idea that these two fluxes measure the same particle population observed on different points of the drift shell.

For all the components of flux, positrons are more abundant than electrons, with a more pronounced abundance in the QT flux inside and outside SAA, compared to the ST one.

This supports a picture where also in the region inside the SAA the dominant processes are the $\pi - \mu - e$ and $\pi - 2\gamma - 2ee$, in particular at energies above $\sim$ tens MeV, and the related shower development.

5. Acknowledgments

We gratefully acknowledge our collegues in AMS, in particular V. Choutko. We greatly benefit of the software libraries (UNILIB,SPENVIS) developed in the context of the Trapped Radiation ENvironment Development (TREND) project for ESTEC, and we thank D.Heynderickx for his help.

This work has been partially supported by Italian Space Agency (ASI) under the contract ARS 98/47.

References

1. Akimov V.V. et al, The main parameters of gamma ray telescope GAMMA-I, 20th ICRC, 2, p.320-323, 1987.
2. Alcaraz,J. et al,The AMS Collaboration, Leptons in Near Orbit Phys Letters B 484 2000,p.10-22
3. Averin S.A. et al., High-energy electrons in the earth's radiation belt, Kosmicheskie Issledovaniya, Vol. 26, N0. 2,, 1988, p. 322.
4. Galper A.M. et al, Discovery of high energy electrons in the radiation belt by devices with gas cherenkov counters, NIM A248 (1986), p. 238.
5. Galper A.M. et al, Electrons with energy exceeding 10Mev in the Earth's radiation belt , in "Radiation Belts: models and standards, Geophys. Monogr, 1997, Lemaire, Heynderickx, Baker editors, Aug. 1997.
6. Getselev I.V. et al., Model of spatial-energetic distribution of charged particles (protons and electrons) fluxes in the Earth's radiation belts, INP MSU Preprint MGU-91-37/241, 1991 (in Russian)
7. E.Fiandrini, G. esposito *et al*, Leptons with E$\geq$200 MeV trapped in Earth's radiation belts, astrp-ph 0106241, accepted for publication on JGR
8. Heynderickx D. et al. Radiation Belts: models and standards, Geophys. Monogr, 1997, Lemaire, Heynderickx, Baker editors, Aug. 1997.
9. Hilton H. , L parameter: a new approximation, J. Geophys. Res., 28, 6952, 1971.

10. Kurnosova, L. V. et al., Flux of electrons above 100 MeV in the earth's radiation belts, Kosmicheskie Issledovaniya, Vol. 79, No. 5, p. 711, 1991.
11. IGRF, http://nssdc.gsfc.nasa.gov/space/model/magnetos/igrf.html
12. McIlwain C. E., Coordinate for mapping the distribution of magnetically trapped particles, J. Geophys. Res., 66, 3681, 1961.
13. Tsyganenko N. A., Determination of Magnetospheric Model Planet Space Sci. 30, 1982.
14. Vette J. I., The AE8 Trapped electron model environment, NSSDC/WDC-A-R&S 91-24, 1991.
15. Voronov S. A. et al., High-energy E: electrons and positrons in the earth's radiation belt, Geomag. and Aeron. Vol. 27, No. 3, p. 424, 1987.
16. TREND project, http://www.magnet.oma.be/home/trend/trend.html, http://www.magnet.oma.be/unilib.html, http://www.spenvis.oma.be/spenvis

SIMULATION OF ATMOSPHERIC SECONDARY HADRON AND LEPTON FLUX FROM SATELLITE TO UNDERGROUND EXPERIMENTS

M. BUÉNERD

Institut des Sciences Nucléaires, IN2P3/CNRS,
53 av. des Martyrs, 38026 Grenoble cedex, France

The successful simulation results obtained in the interpretation of the recent AMS measurements of proton, leptons and light nuclei particle flux, are reviewed. A similar success is being met in the analysis of secondary particle flux measured in the atmosphere (antiprotons, muons) and in underground experiments (neutrinos).

Keywords: Cosmic Rays; Atmospheric secondaries; Hadroproduction.

1. INTRODUCTION

Recent experiments have reported new measurements of atmospheric secondary particle flux over a broad range of altitudes, depending on the particular species, extending from satellite and balloon altitudes, down to ground and underground experiments. This presentation reviews the results obtained recently in a simulation work undertaken to account for the measured flux of secondary particles in terms of dynamical interactions between Cosmic Rays (CR) and the atmosphere.

The primary purpose of the work was to understand the proton, $e^{\pm}$, and light ion flux measured recently by the AMS experiment [1,2,3,4]. However, while this undertaking was considered, it was also felt that it could contribute fruitfully to various other experiments involving directly or indirectly CR-atmosphere interactions like $\bar{p}$, and lepton flux measurements in balloon experiments, and the underground neutrino experiments, as well as to the evaluation of the secondary particle populations for the preparation of future satellite experiments like AMS2, PAMELA, etc...

The results presented in the following have been reported in scientific journals [5,6,7], or conference proceedings [8,9,10,11], while some others are still preliminary and will be reported in forthcoming publications [13,14].

The first part of the talk will briefly describe the general features of the simulation program. In the second part, the results obtained in the interpretation of the AMS data on the p, $e^{\pm}$, d and ^3He, flux will be reviewed. The third part will be devoted to the simulation of the secondary $\bar{p}$ and μ flux in the atmosphere, and to the atmospheric neutrino flux in underground experiments.

2. The simulation program

The basic assumptions on the physics processes governing the production and the propagation of the flux of secondary charged particles in the terrestrial environment, incorporated in the simulation program were the following:

- The secondary particles are produced by the interaction between the Cosmic Ray flux (CR) and the atmosphere.
- For the present purpose, the CR flux can be limited to a good approximation to its main components: proton and helium. These flux are described in accordance with the recent AMS measurements [1,2,4] with appropriate corrections for solar modulation effects made.
- The density of the atmosphere is described according to the currently available model of atmosphere.
- All charged particles propagate in the earth magnetic field described by a multipole development on the geomagnetic angular coordinates.
- For each secondary particle species, particles are generated according to their specific inclusive production differential cross sections, which energy and angular dependence is parametrized as described below.
- Secondary particle propagation and interactions are processed the same way as for incoming CRs, leading to the development of atmospheric cascades involving up to about 10 generations of successive collisions. All charged particles undergo energy loss by ionization.

This approach therefore takes into account the particle productions in secondary interactions at all stages inside the atmospheric cascade. This has quite significant effects as it will be seen below.

(For technical and bibliographical details on the above, see [5,6]).

2.1. Production cross section for secondary particles

With the incoming flux of proton and helium as a common feature to all calculations, the inclusive triple differential cross sections appropriate for each production channel have to be calculated separately in the event generator. The following production cross sections were used for the calculations of the various secondary particle flux studied:

2.1.1. Protons

Protons are produced in $CR + A \rightarrow p + X$ type reactions, with CR, A, and X standing for CR particles $(p, ^4He)$, atmospheric nuclei (weighted average value of atmosphere components), and recoil system, respectively [5]. The calculated cross section included two components corresponding to quasi free (forward) and deep inelastic (backward) scattering processes [18], respectively.

2.1.2. Light nuclei

Light nuclei were assumed to originate from $CR + A \to a + X$ reactions with $a =^{2,3}H$, $^{3,4}He$. Two models were considered and implemented to describe this production, corresponding to different final state kinematics: The fragmentation model and the coalescence model, both in their simplest versions [7].

2.1.3. Leptons

This is a main aspect of the work since several lepton flux have been measured, in balloon, satellite, and underground experiments, which all originate from the same reaction type, at least for the dominant mechanism. The flux calculations for these species are then most tightly, and thus sensitively, constrained by the data. The main mechanism of lepton generation considered was through the pion production reaction chain $CR + A \to \pi + X$, $\pi \to \mu + \nu$, $\mu \to e + \nu + \bar{\nu}$, with CR and A defined as above. The kaon contribution is typically one order of magnitude smaller; It was considered only for the neutrino and muon flux calculations [14]. These cross sections are governing the populations of $e^{\pm}, \mu^{\pm}, \nu$ produced in the atmosphere, and it is one of the major successes of the works presented here, to reproduce simultaneously the measured flux of $e^{\pm}$ at the (satellite) altitude of AMS (see also [9] for the $e^{\pm}$ flux in the atmosphere), of $\mu^{\pm}$ in the atmosphere from ground level to TOA, and the Est-West neutrino asymmetry measured by superK, to a fair accuracy with no adjustable parameter. The inclusive pion differential cross section used in the calculations was based on a modified version of the functional form proposed in [16], and fitted to a broad range of pion data [17]. Good fits to the data could be obtained through the covered momentum range ($\approx$1.5-450 GeV/c). For the Kaon production, the form and parameters from [16] were used. For the electron flux, the (tertiary) pair production cross section $\gamma + A \to e^+ e^- + A$ induced by Bremstrahlung photons and π^0 decay photons, was also calculated [6].

2.1.4. Antiprotons

For the antiprotons, a modified form based on the Ref [16] parametrized functional was used for the description of the inclusive production reaction $CR + A \to \bar{p} + X$. It was fitted on a set of antiproton data [12] measured at incident energies between about 12 and 24 GeV, for which good fits could be obtained. The work is being extended to higher energies with good results up to 300 GeV in the lab frame.

3. Interpretation of the AMS results

3.1. Protons

The first interpretation effort of the AMS data using the present approach, was devoted to the strong component of protons ("second spectrum" in [1]) below the geomagnetic cutoff (GC). The results are illustrated on fig 1. The data are very well

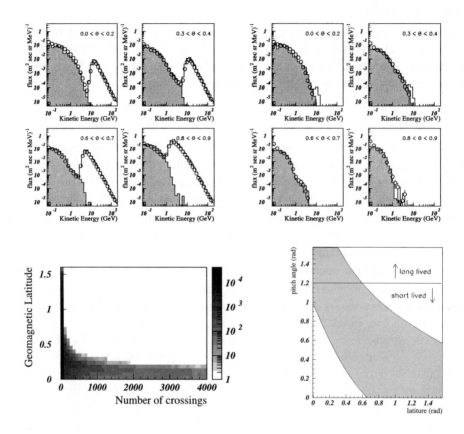

Fig. 1. Top: Experimental upward and downward proton flux measured by AMS compared with the simulation results. Bottom left: Simulated flux of long-lived (trapped and semi trapped) protons. Bottom right: AMS pitch angle acceptance versus latitude, explaining the confinement of the trapped protons in the equator region as seen on the left. From Refs.[5,15]

reproduced by the calculations for all latitudes, including the strong flux enhancement in the GeV region of the spectrum for equatorial latitude. The latter appears to be due to trapped or semi-trapped (long lived) particles. Bottom left Fig. 1 shows the scatter plot of the long-lived population of the simulated sample, which appears to be confined in the equatorial region. The explanation for this feature is given on the right hand panel [15] which shows the particle pitch angle (PA) at the AMS location versus geomagnetic latitude plane. In this plane, trapped particles are allowed only in the area above the horizontal line, i.e. for PA>≈1.2, where mirror points of particle trajectories lie outside (above) the atmosphere. Below this line, they are inside the atmosphere, where particles are dumped and thus cannot be trapped. Particle directions have also to be within the AMS acceptance shown by the broad grey stripe going from the upper left to the lower right across the figure. The area of the plane allowed to trapped particles is thus the upper left corner (large pitch angles and low latitudes) limited on one hand by the mirror point condition and on the other hand by the AMS acceptance. The corresponding upper limit in latitude (≈ 0.6 rad) is seen to be in agreement with the latitude distribution of the simulated population of long-lived protons (left panel).

3.2. *Electrons and positrons*

The flux of e^+ and e^- particles measured by AMS was surprising by two respects: Firstly, it displayed a similar strong second spectrum below GC as observed for protons; Secondly, The e^+ over e^- flux ratio evolved from about four around the equator down to about one in the polar region [3]. These results are quite well reproduced by the calculations, as seen on fig 2. The analysis of the simulation data showed that the strong charge asymmetry observed for equatorial latitudes is due to a combined effect of the East-West (EW) dependence of the GC, with the forward peaking of the lepton production reactions, and the absorption effect of the

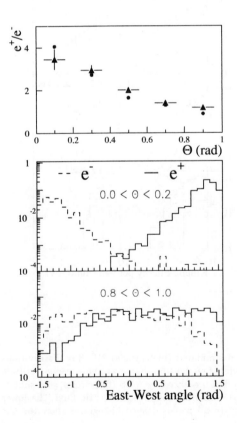

Fig. 2. Top: Experimental (full circles) and simulated (triangle) lepton asymmetry versus geomagnetic latitude. Middle: East-West angle distributions of e^+ and e^- showing the large EW asymmetry for equatorial latitudes. Bottom: Same for polar region showing as expected, a vanishing asymmetry. From Ref.[6]

atmosphere [6].

3.3. Light ions

The AMS data surprisingly uncovered the existence of a small but sizeable subGC component of ^3He, whereas no ^4He flux could be detected (above some upper limits set by the experimental conditions) [4], while the latter is dominating the former in the CR flux by a sound order of magnitude. This apparent paradox found a natural explanation in the framework of the nuclear coalescence model which allowed to account quantitatively for all these features with a good accuracy [7]. This is illustrated on fig 3 left, where the experimental energy spectrum and latitude distribution of ^{3}He particles are seen to be well reproduced by the calculations. The deuterium flux below GC is also very well reproduced by the calculations using the same model (fig 3 right). This consistent agreement between data and calculations (no adjustable parameter), for two populations of light nuclei provides a strong support to the validity of the theoretical basis of the approach.

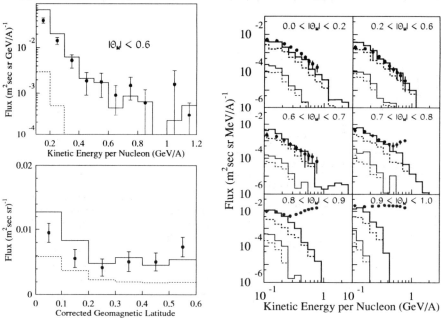

Fig. 3. Left top: Experimental (full circles) and simulated (histograms) ^{3}He flux distributions below GC, versus energy (integrated over latitude). Left bottom: Same as above versus latitude (integrated over energy). Right: Deuterium (^{2}H) flux measured by AMS (full circles) and subGC component calculated by simulation (histogram), simultaneously with the ^{3}He flux. The lower histograms give the calculated tritium (^{3}H) flux. In all panels dashed histograms show the CR He induced contributions. From Ref.[7]

4. Secondary atmospheric particle flux

The successful interpretation of the AMS data opened new prospects for calculating the flux of secondary atmospheric particles like antiprotons, muons, and neutrinos relevant to several ongoing experimental programs, and to future new experiments.

4.1. *Antiprotons*

The BESS [19] and CAPRICE [21] experiments are studying the antiproton CR flux in high altitude balloon experiments. In these experiments, the raw data have to be corrected for the atmospheric $\bar{p}$ production to obtain the galactic flux. Fig. 4 shows the CR $\bar{p}$ flux obtained by BESS, together with the applied corrections (curve), and those calculated in the present work (histogram). The flux expected from the simulation results apppears to be significantly larger than the values obtained from the transport equation approach. This is observed for both the BESS (left) and CAPRICE (middle) data. Fig 4 right shows the preliminary results of measurements performed at terrestrial altitude (2770 m) by the BESS collaboration [20] compared to the preliminary simulation results at this altitude [13]. The agreement is excellent and consistent with the previously quoted results, showing that the particle production in the atmosphere seems to be fairly well understood and the underlying dynamics and kinematics appropriately treated. It must be pointed out however that the calculated proton flux at this altitude significantly overestimate the proton data from the same experiment, a point of disagreement currently under examination. At sea level, the antiproton flux in the GeV range is calculated to be of the order of 2 $[m^2/sr/GeV/c/mn]^{-1}$. A small flux, but large enough to allow the testing of the $\bar{p}$ identification capability of space detectors on the ground [11,13].

The antineutron flux at terrestrial altitudes is being calculated in the same

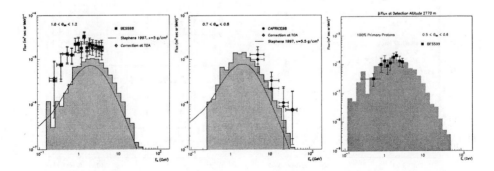

Fig. 4. Experimental $\bar{p}$ flux (symbols) measured in the atmosphere. Left: Low energy BESS measurements at high altitude; Middle: high energy CAPRICE data; Right: 2770 m BESS measurements (preliminary [20]). The data are compared to (preliminary) simulation results (histogram). The curves show the calculated corrections (atmospheric flux) to the raw flux in the original works (see text).

conditions in the perspective of an experimental project [22].

4.2. Muons

The muon flux measurements together with the $e^{\pm}$ measurements constitute the most constraining body of data currently available for a stringent test of any theoretical or phenomenological approach of the atmospheric production of leptons. This is of major importance regarding the atmospheric neutrino issue in relation with the neutrino oscillations data which interpretation relies to some extent on calculations [25] and of which some results are still awaiting for a quantitative interpretation (see below) [14]. The subject has been addressed in several previous works using various theoretical approaches (see [14] for the context).

Fig 5 shows a set of μ^- spectra measured recently by the CAPRICE [23] and HEAT [24] experiments for altitudes ranging from ground level up to 38 km, compared with the simulation results [14]. The right panel shows the ratio of calculated to measured values. The agreement between calculations and data is very good through the range of altitudes, but for the 38 km spectrum for which the calculations significantly underestimate the experimental flux for low energy muons below about 1 GeV. This defect seems to be related to the low energy pion yield of the event generator. It has turned out in this study, that the currently existing pion

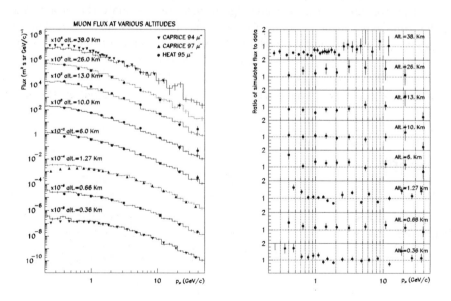

Fig. 5. Left: Experimental (full circles) μ^- flux measured by the BESS and CAPRICE experiments compared with simulated values (histograms, preliminary results). Right: Ratio of the experimental to simulated values.

data are not enough constraining for the relevant cross section parameters to be firmly established. Consequently, the low energy π yield cannot be calculated with a good enough accuracy for the μ yield to be reproduced accurately. However, this inaccuracy has a limited influence on the calculated neutrino flux [14].

4.3. *Atmospheric neutrinos*

After the $e^{\pm}$ and $\mu^{\pm}$ have been calculated and shown to reproduce the measured data with a good enough accuracy, the last stage of the approach is the calculation of the neutrino flux. The whole the range of geomagnetic coordinates have been spanned in the calculations, and the four neutrino energy spectra as well as the zenith and azimuth distributions have been calculated as a function of the latitude and longitude coordinates [14]. In addition, the flux have been calculated at the locations of the existing underground experiments. They have been used to calculate the East-West asymmetries A_{EW} for the e-like and μ-like events in the same conditions as reported by the SuperK experiment [26], i.e. same cuts on the energy range, and compared to the experimental values. In addition, in the calculations, the e and μ production angles have been approximately taken into account by using a functional form adjusted on the values reported in [26]. The preliminary results for the mean values of A_{EW} calculated in the above conditions for the two class of events are in quite good agreement with the superK measurements. The experiment reported: $A_{EW}^{e-like}(exp)=0.21\pm0.04$ and $A_{EW}^{\mu-like}(exp)=0.08\pm0.04$, to be compared to the values obtained in the simulation: $A_{EW}^{e-like}(sim)=0.27\pm0.1$ and $A_{EW}^{\mu-like}(sim)=0.092\pm0.05$ [10,14]. Not taking into account the lepton production angles leads to (neutrino) EW asymmetry pretty far from these values.

5. Summary and outlook

In summary, it has been shown that the simulation model described in this presentation allows to successfully account for the charged particle flux measured by AMS01 below the geomagnetic cutoff at satellite altitudes, and for the hadron (p, $\bar{p}$) and lepton ($e^{\pm}$, $\mu^{\pm}$) flux measured in the atmosphere down to ground level, as well as in the underground superK neutrino experiment. Final calculations are currently being processed for atmospheric antiprotons and leptons.

Finally, the numerical tool developed for these analysis should also provide a powerful mean of estimating the atmospheric flux of secondary particles for future satellite experiments like AMS02, PAMELA, GLAST, etc..., as well as for currently running balloon and underground (superK, SOUDAN) experimental programs. It is also clearly applicable to the investigation of other dynamical ranges of CR-atmosphere interactions, such as the mechanism and the dynamics of population of the radiation belts in a much lower energy domain, and likely to other similar topics of geophysical interest.

Acknowledgements

The author is indebted to his collaborators, L. Derome, Yong Liu, B. Baret, and C.Y. Huang, for their help in the preparation of this talk.

References

1. AMS Collaboration, J. Alcaraz et al., Phys. Lett. **B 472**, 215(2000).
2. AMS Collaboration, J. Alcaraz et al., Phys. Lett.**B 490**, 27(2000)
3. AMS Collaboration, J. Alcaraz et al., Phys. Lett. **B 484**, 10(2000).
4. AMS Collaboration, J. Alcaraz et al., Phys. Lett.**B 494**, 193(2000).
5. L. Derome et al., Phys. Lett. **B 489**, 1(2000); L. Derome, and M. Buénerd, Nucl. Phys. **A 688**, 66c(2001).
6. L. Derome, M. Buénerd, and Yong Liu, Phys. Lett. **B 515**, 1(2001).
7. L. Derome, and M. Buénerd, Phys. Lett. **B 521**, 139(2001).
8. L. Derome and M. Buénerd, *27th ICRC, Hamburg, Aug. 7-14, 2001, OG1.019-20*.
9. L. Derome, Yong Liu, and M. Buénerd, ref [8], OG1.018
10. Yong Liu, L. Derome, and M. Buénerd, ref [8], OG2.02
11. C.Y. Huang, L. Derome, and M. Buénerd, ref [8], OG1.014
12. C.Y. Huang and M. Buénerd, *Parametrization of the inclusive antiproton cross section produced by 11 to 24 GeV protons on nuclei*, Report No. ISN-01/18, March 2001 (unpublished).
13. C.Y. Huang, M. Buénerd, and L. Derome, in preparation; C.Y. Huang, PhD thesis, Université J. Fourier, Grenoble, in preparation.
14. Yong Liu, L. Derome, and M. Buénerd, *Atmospheric muon and neutrino flux from 3-dimensional simulations*, In preparation.
15. L. Derome, private communication.
16. A. N. Kalinovsky, N. V. Mokhov, and Y. P. Nikitin, *"Passage of High Energy Particles Through Matter"*, New-York, AIP ed., 1989.
17. Yong Liu, L. Derome, and M. Buénerd, *Parametrization of the inclusive production cross section of proton induced $\pi^{\pm}$ particles on nuclei*, ISN Internal Report 01-012, Feb 2001 (unpublished).
18. Y.D. Bayukov et al., Sov. J. Nuc. Phys. **42**, 116(1985).
19. T. Maeno et al., AstroPart. Phys. **16**, 121(2001)
20. M. Fujikawa, PhD thesis, University of Tokyo, 2001; M. Nozaki, private communication; and BESS collaboration, to be published.
21. M. Boezio et al., ApJ **561**, 787(2001)
22. A. Menchaca-Rocha et al., Private communication.
23. J. Kremer et al., Phys. Rev. Lett. **83**, 4241(1999)
24. S. Coutu et al., Phys. Rev. **D 62**, 032001(2000)
25. T. Kajita and Y. Titskuba, Rev. Mod. Phys.**73**, 85(2001)
26. Y. Fukuda et al., Phys. Rev. Lett. **81**, 1562(1998)
27. T. Fugutami et al., Phys. Rev. Lett. **82**, 5194(1999)

REVIEW ON MODELLING OF THE RADIATION BELTS

D. HEYNDERICKX

Belgisch Instituut voor Ruimte-Aëronomie
Ringlaan 3, B-1180 Brussel, Belgium

The Earth's trapped radiation belts were discovered at the beginning of the space age and were immediately recognised as a considerable hazard to space missions. Consequently, considerable effort was invested in building models of the trapped proton and electron populations, culminating in the NASA AP-8 and AE-8 models which have been the de facto standards since the seventies. The CRRES mission has demonstrated that the trapped radiation environment is much more complex than the static environment described by the old models. Spatial and especially temporal variations were shown to be much more important than previously thought, and to require more complex models than those in use at that time. Such models are now becoming available, but they are limited in spatial or temporal coverage, and no global, dynamic, trapped radiation belt model is forthcoming. It is therefore vital to co-ordinate future modelling efforts in order to develop new standard models.

Keywords: radiation belts; radiation effects; modelling.

1. Introduction

This paper presents a brief review of the main characteristics of the Earth's trapped radiation belts and of the engineering models that are used to evaluate mission fluences and doses. More detailed descriptions of the Earth's radiation environment and the related physical processes can be found in Refs. 1-3.

The Belgisch Instituut voor Ruimte-Aëronomie (BIRA) has developed for the European Space Agency (ESA) a World-Wide Web interface to models of the space environment and its effects on spacecraft and systems. The SPace ENVironment Information System (SPENVIS) can be accessed at http://www.spenvis.oma.be/spenvis/. The figures in this paper, except those with a reference number, were produced with the SPENVIS plotting facilities.

2. The Earth's trapped radiation environment

2.1. *The concept of trapped radiation*

The motions of charged particles entering the magnetosphere from the solar wind and undergoing acceleration, or resulting from the decay of neutrons produced by cosmic ray interactions with the neutral atmosphere, are dominated by the magnetospheric magnetic field.

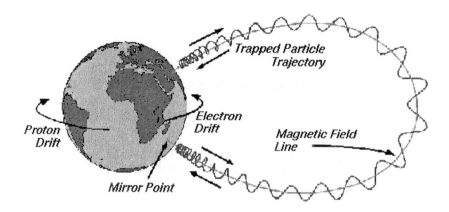

Fig. 1. Illustration of the motion of trapped particles in the Earth's magnetic field

The motion of these energetic charged particles consists of three components (see Fig. 1):

(1) gyration about magnetic field lines;
(2) movement of the gyration centre up and down magnetic field lines (guiding centre motion);
(3) slow longitudinal drift of the guiding centre path around the Earth, westward for ions and eastward for electrons.

The resulting trajectories lie on toroidal surfaces, called drift shells, centred on the Earth's dipole centre. Particles confined to a drift shell can remain there for long periods, up to years for protons at altitudes of a few thousand kilometers, whence the term "trapped particles".

The population of charged particles stably trapped by the Earth's magnetic field consists mainly of protons with energies between 100 keV and several hundred MeV and electrons with energies between a few tens of keV and 10 MeV. There is also evidence for the existence of a narrow region centred around altitudes of about one Earth radius containing trapped heavy ions which are believed to be decelerated anomalous cosmic ray ions; the intensities of these heavy ions are several orders of magnitude below the intensities of trapped energetic protons in this region.

2.2. *The trapped proton population*

The energetic (above 10 MeV) trapped proton population is confined to altitudes below 20,000 km, while lower energy protons cover a wider region, with protons below 1 MeV reaching geosynchronous altitudes. Figure 2 shows the distribution of trapped protons with energies above 10 MeV, as predicted by the NASA AP-8 MAX model[4], in invariant coordinate space. The region of space covered by higher energy protons diminishes with increasing energies and the location of the highest

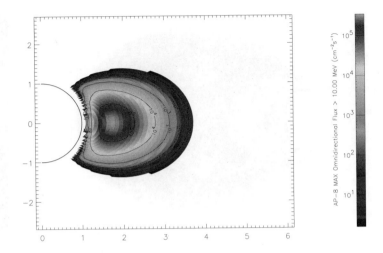

Fig. 2. Invariant coordinate map of the AP-8 MAX integral proton flux > 10 MeV. The semicircle represents the surface of the Earth, distances are expressed in Earth radii.

intensities moves inward.

2.3. *The trapped electron population*

Figure 3 shows the AE-8 MAX[5] trapped electron population above 1 MeV in invariant coordinate space. The population distribution is characterised by two zones of high intensities, below altitudes of one Earth radius and above two Earth radii in the magnetic equatorial plane, respectively, which are separated by a region of low intensities, called the slot region. The location and extent of the inner and outer belts and of the slot region depends on electron energy, with higher energy electrons confined more to the inner belt, and lower energy electrons populating the outer belt to altitudes beyond geosynchronous orbit. Note that at high latitudes the outer electron belt reaches down to very low altitudes.

2.4. *Dynamics of the trapped particle population*

The general description of the radiation belts in Sections 2.2 and 2.3 represents what could be called the average particle distributions based on the static NASA models AP-8 and AE-8[6]. However, it has long been established that the actual population is very dynamic over different time scales.

2.4.1. *Solar cycle effects*

The variation of solar irradiance with the 11-year solar cycle induces a periodicity of the low altitude trapped proton and electron fluxes: during solar maximum

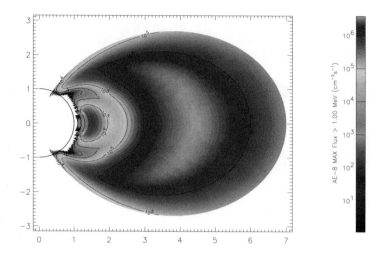

Fig. 3. Invariant coordinate map (see Fig. 2) of the AE-8 MAX integral electron flux > 1 MeV.

the Earth's neutral atmosphere expands compared to solar minimum conditions, so that the low altitude edges of the radiation belts are eroded due to increased interactions with neutral constituents. Figure 4 shows the variation of the low altitude trapped proton flux over the solar cycle[7]. The erosion effect increases with decreasing altitude and the recovery of the population shows a phase lag which also depends on altitude.

2.4.2. *Secular changes in the geomagnetic field*

The low altitude trapped particle population is also influenced by secular changes in the geomagnetic field[8]: the location of the centre of the geomagnetic dipole field drifts away from the centre of the Earth at a rate of about 2.5 km/year (the separation currently exceeds 500 km), and the magnetic moment decreases with time. The combined effect is a slow inward drift of the innermost regions of the radiation belts.

The separation of the dipole centre from the Earth's centre and the inclination of the magnetic axis with respect to the rotation axis produce a local depression in the low altitude magnetic field distribution at constant altitude. As the trapped particle population is tied to the magnetic field, the lowest altitude radiation environment (below about 1,000 km) peaks in the region where the magnetic field is depressed[1]. This region is located to the south east of Brasil, and is called the South Atlantic Anomaly (SAA). Figures 5 and 6 represent a world map at 500 km altitude of the trapped proton (> 10 MeV) and trapped electron (> 1 MeV) distributions, respectively. The SAA shows up clearly in both maps. Proton fluxes are negligible

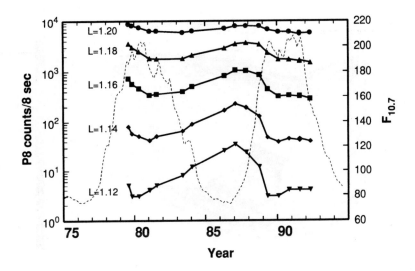

Fig. 4. Variation of proton count rates in the 80–215 MeV channel of the MEPED detector aboard the TIROS/NOAA spacecraft over the solar cycle as a function of L^7. The dashed line shows the 13-month smoothed solar $F_{10.7}$ flux.

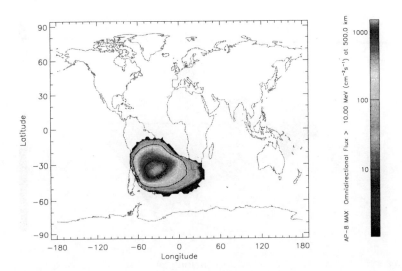

Fig. 5. World map of the AP-8 MAX integral proton flux > 10 MeV at 500 km altitude.

outside the SAA, but electron fluxes can be very high at high latitudes where field lines from the outer electron belt reach down to low altitudes. A further effect of the secular change in the geomagnetic field is a slow westward drift of the SAA at a rate of 0.3 deg per year[9].

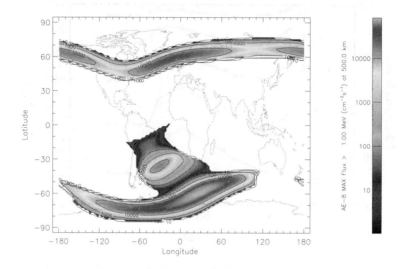

Fig. 6. World map of the AE-8 MAX integral electron flux > 1 MeV at 500 km altitude.

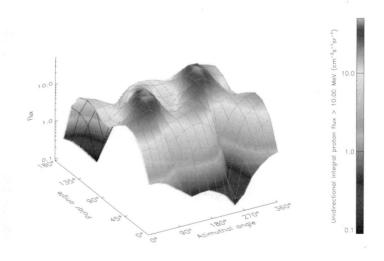

Fig. 7. Angular dependence of the AP-8 MAX integral proton flux > 10 MeV, averaged over an 800 km geosynchronous orbit. Angles are measured in a reference frame with its polar axis parallel to the satellite velocity vector.

2.4.3. *Low altitude trapped proton anisotropy*

At low altitudes (typically below 2,000 km), trapped particles interact with the neutral atmosphere. The gyroradii of trapped protons with energies above 1 MeV

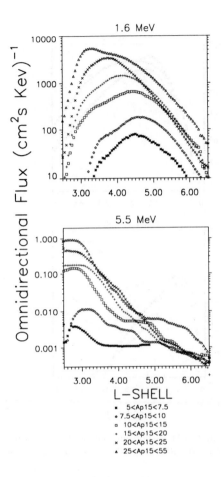

Fig. 8. Profiles of 1.6 MeV (top) and 5.5 MeV (bottom) omnidirectional electron flux on the magnetic equator, as a function of L, taken from the CRRES electron models for six ranges of A_{p15}[11].

are comparable to the atmospheric scale height, which means that during a gyration motion they encounter different atmospheric densities. As a result, proton fluxes depend on their arrival direction in the plane perpendicular to the local magnetic field vector (as well as on their pitch angle). The resulting anisotropy is called the East-West effect, and can cause differences of a factor three or more in fluxes arriving from different azimuths. The effect is illustrated in Fig. 7, which shows the angular dependence of the AP-8 MAX integral proton flux > 10 MeV, averaged over an 800 km geosynchronous orbit.

2.4.4. Magnetospheric conditions

Besides the long term variations in the trapped particle population described in Sections 2.4.1–2.4.2, variations on much shorter time scales occur as well.

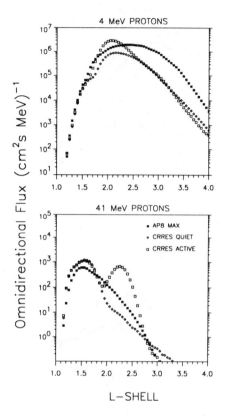

Fig. 9. Profiles of 4 MeV (top) and 41 MeV (bottom) omnidirectional proton flux on the magnetic equator, as a function of L, obtained with the CRRES quiet and active proton models and with AP-8 MAX[11].

Outer zone electrons can vary in intensity by orders of magnitude over periods of a few hours. Measurements with instruments onboard the Combined Release and Radiation Effects Satellite (CRRES) have shown that there are also major changes in the spatial distributions of outer zone electrons[10]. Gussenhoven et al.[11] have shown that the changes in flux and spatial distribution can be ordered by level of magnetospheric activity, i.c. the fifteen day running average of A_p. Figure 8 shows omnidirectional electron flux profiles on the magnetic equator as a function of McIlwain's[12] L for six ranges of A_p15.

CRRES Data also demonstrated that magnetic storms can greatly influence the trapped proton population[13]. The March 1991 storm created a second, stable high energy belt above $L=1.8$, with peak flux values exceeding pre-storm values by an order of magnitude[11], as shown in Fig. 9. The newly-created proton belt decayed only very slowly and was still present six months later when the CRRES satellite was lost.

3. Effects of trapped radiation on spacecraft and components

Due to their large energy coverage, trapped particles cause a variety of effects in spacecraft, components and biological systems.

Low energy electrons contribute to spacecraft surface charging. High energy electrons injected and accelerated through the magnetotail can cause dielectric charge buildup deep inside geosynchronous spacecraft which may lead in turn to destructive arcing. Inner and outer belt electrons also contribute to ionising doses through direct energy deposition and bremsstrahlung effects.

High energy protons in the inner radiation belt are the main contributors to ionising dose deposition in shielded components. They also dominate Single Event Upset (SEU) rates at low altitudes and latitudes, where cosmic rays and solar energetic particles are effectively shielded by the geomagnetic field. Lower energy protons (up to 10 MeV) contribute to Non-Ionising Energy Loss (NIEL) dose which affects Charged-Coupled Devices (CCD) and other detectors; unshielded detectors can be affected even in the outer belt, where < 1 MeV protons are present.

4. Shortcomings of present day radiation belt models

The old NASA AP-8 and AE-8 radiation belt models[6] are still the de facto standards for engineering applications. This is mainly due to the fact that up to now they are the only models that completely cover the region of the radiation belts, and have a wide energy range for both protons and electrons. It should be noted that a considerable part of the range of the NASA models was achieved by extrapolation.

The NASA models are static and are in principle only valid for the period when the data for the models were obtained[16]. In view of the dynamic characteristics of the radiation belts outlined in Section 2.4, it is clear that correspondingly dynamic models are needed for accurate predictions of mission fluences and doses. Several efforts are under way to include dynamic behaviour in new radiation belt models, but up to now no models are available that duplicate the spatial and energy range of the NASA models. In order to achieve this, high quality data are needed from different locations in the magnetosphere, covering long time periods and with high resolution in energy, direction and time. Simple radiation monitors could easily be installed on commercial satellites, which would help the continuous upgrading needed for truly dynamic radiation belt models. However, as long as not all features of the radiation belts are fully understood or adequately modelled, high quality data are indispensable.

References

1. J. G. Roederer, *Dynamics of Geomagnetically Trapped Radiation* (Springer-Verlag, 1970).
2. M. Walt, *Introduction to Geomagnetically Trapped Radiation* (University Press, Cambridge, 1994).

3. J. F. Lemaire, D. Heynderickx and D. N. Baker (eds), *Radiation Belts. Models and Standards, Geophysical Monograph* **97** (AGU, 1996).
4. D. M. Sawyer and J. I. Vette, NSSDC/WDC-A-R&S 76-06, 1976.
5. J. I. Vette, NSSDC/WDC-A-R&S 91-24, 1991.
6. J. I. Vette, NSSDC/WDC-A-R&S 91-29, 1991.
7. S. L. Huston, G. A. Kuck and K. A. Pfitzer, in *Radiation Belts. Models and Standards*, ed. J. F. Lemaire, D. Heynderickx, and D. N. Baker, *Geophysical Monograph* **97**, p. 119 (AGU, 1996).
8. A. C. Fraser-Smith, *Rev. Geophys.* **25**, 1 (1987).
9. D. Heynderickx, *Nucl. Tracks Radiat. Meas.* **26**, 369 (1996).
10. M. S. Gussenhoven, E .G. Mullen and D. H. Brautigam, *IEEE Trans. Nucl. Sci.* **43**, 353 (1996).
11. M. S. Gussenhoven, E. G. Mullen and D. H. Brautigam, in *Radiation Belts. Models and Standards*, ed. J. F. Lemaire, D. Heynderickx, and D. N. Baker, *Geophysical Monograph* **97**, p. 93 (AGU, 1996).
12. C. E. McIlwain, *J. Geophys. Res.* **66**, 3681 (1961).
13. M. K. Hudson, S. R. Elkington, J. G. Lyon, V. A. Marchenko, I. Roth, M. Temerin and M. S. Gussenhoven, in *Radiation Belts. Models and Standards*, ed. J. F. Lemaire, D. Heynderickx, and D. N. Baker, *Geophysical Monograph* **97**, p. 57 (AGU, 1996).
14. S. M. Seltzer, NBS Technical Note 116, 1980.
15. J. Feynman, G. Spitale, J. Wang and S. Gabriel, *J. Geophys. Res.* **98**, 13,281 (1993).
16. D. Heynderickx, J. Lemaire and E. J. Daly, *Nucl. Tracks Radiat. Meas.* **26**, 325 (1996).

LOW ENERGY SOLAR AND GALACTIC COSMIC RAYS AT 1 AU

MARCO CASOLINO*

INFN Roma2 and University of Roma Tor Vergata
Via della ricerca scientifica 1, 00133 Roma, Italy

In this work we present some of the recent results obtained with the satellite missions NINA-1 and NINA-2 and the experiments on board MIR space station Sileye-1 and 2. The aim is the study of the low energy (10 MeV - 2 GeV) cosmic ray component and different periods of the solar cycle and during Solar Energetic Particle events. Other items of physics include the measurement of the secondary cosmic ray component, produced in the interaction with the upper layers of Earth's atmosphere and the evaluation of the absorbed and equivalent doses inside MIR.

Keywords: Cosmic rays, Solar Energetic Particles, Antimatter component of Cosmic rays.

1. Introduction

Cosmic rays are a versatile tool to gather information for the comprehension of several physical processes. Their importance comes from their wide energy range (20 orders of magnitude) and rich nature (hadrons: nuclei, antiprotons; leptons: $e^{\pm}$, ν; photons). They are the product of phenomena with range scaling from cosmological issues (i.e. the antimatter component of c.r.) to extragalactic and galactic phenomena (i.e. gamma ray bursts, supernovae), the Sun-Earth environment (Solar Particle Events, magnetohydrodinamics) and the radiation environment in space and on aircrafts. A number of very different detectors is used to study and approach the various problems posed: we go from the underground detectors to the large array telescopes to the balloon and satellite experiments. Satellite and probes carry different detectors on the same spacecraft to cover wider energy and particle range; the study of cosmic rays within the heliosphere requires a number of different probes in different points of the solar system to provide a picture of the spatial and temporal variations driven by our Sun of cosmic rays of galactic and solar nature.

In this vast context we focus in this work on some of the recent results obtained with the NINA-1 (1998-1999) and NINA-2 (2000-2001) satellite missions and the Sileye-1 (1995-1997) and Sileye-2 (1997-2000) detectors on board the Mir space station. These experiments are focused upon the study of the low energy ($\simeq 10 MeV - 2 GeV$) component of cosmic ray nuclei (from p to Fe) within Earth's magnetosphere. Their

---

*E-mail: casolino@roma2.infn.it

position and orbit puts them in a particularly suitable point to address not only galactic cosmic ray physics issues but also solar and magnetospheric effects related to the interaction of cosmic rays with the geomagnetic field. Thanks to the success of these missions, this program will continue and be expanded in the years following the 23rd solar maximum with the Pamela experiment, primarily devoted to the search of antimatter in space and to low energy physics, and the Sileye-3 (Alteino) experiment, devoted to cosmic ray studies in the International Space Station.

2. NINA-1 and -2

The NINA (*New Instrument for Nuclear Analysis*) detectors have been developed by a joint program of the Italian National Institute of Nuclear Physics (INFN) and the Moscow Engineering and Physics Institute (MEPhI). This is a part of a wider collaboration, the WiZard group, which has carried out balloon-borne experiments for the detection of cosmic antiparticles since 1989[1]. NINA-1 was mounted on board the Russian Resurs-O1 n.4 satellite and launched on July 10th, 1998; it was operational up to the middle of the year 1999 when transmissions stopped due to a failure to the telemetry system. NINA-2 was installed as scientific payload of the first ASI (Italian Space Agency) satellite MITA and launched on July 15th, 2000[2]; it was operational until August 2001 when the satellite re-entered the atmosphere in Antarctica. Both devices carry two versions of the same telescope, composed of 16 planes, each made of two n-type silicon detectors, 6×6 cm^2, segmented in 16 strips and orthogonally mounted so to provide the X and Y information of the particle track. The thickness of the first two detectors is 150 μm; all the others are 380 μm thick, for a total of 11.7 mm of active silicon. The 16 planes are vertically stacked. A detailed description of the detector can be found in [3,4]. The high inclination orbit (respectively 98° and 82°) of the satellites and their low altitude (800 km and 300-450 km) allow the study of solar and galactic cosmic rays in the energy range 10–200 MeV/n. The increased telemetry and data handling capabilities of NINA-2 allow to reach an upper energy range of 2 GeV by operating with a less restrictive trigger system. The instruments are capable to identify nuclei up to Fe with isotopic resolution up to C. Passage in the equatorial regions provides information on the trapped and secondary component of cosmic rays; the complementary period of observation (going toward the solar maximum and at solar maximum) allows the study of solar modulation effects.

3. Galactic Cosmic Rays

Cosmic rays are dominated by the proton and helium component ($\simeq$ 90% and $\simeq$ 10% respectively); above $\simeq$ 1 GeV/n cosmic rays are almost exclusively of galactic origin; the differential energy spectrum can be represented by a power law of index 2.68 up to $10^4 GeV$ (the knee) where the index steepens to 3.15. Below 1 GeV the spectrum decreases (due to solar wind particles and magnetic field which act as a potential barrier for interstellar cosmic rays) to reach a minimum at about 10 MeV;

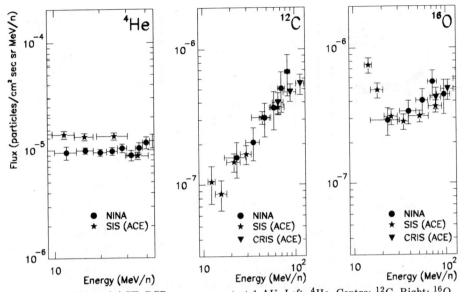

Fig. 1. NINA and ACE GCR measurement at 1 AU: Left: ^{4}He, Centre: ^{12}C, Right: ^{16}O.

below this value the particle rate increases due to the contribution of the solar and anomalous components[6].

Figure 1 shows the differential energy spectra for ^{4}He, ^{12}C, and ^{16}O, measured by NINA in the solar quiet period December 1998-March 1999 selected restricting the data set to high latitude regions (L-shell>6)[a] excluding solar active days. The flatness of the ^4He spectrum is due to the anomalous component, while the increase of Carbon and Oxygen above to 20 MeV is due to the galactic component. Data are compared with SIS detector on ACE, about in the same period of observation. Data from SIS on ACE belong to a cycle of 27 days[4] from 1999 February 6 to 1999 March 4, and are the sum of ^{3}He and ^{4}He. There is a general agreement among the two sets of results; the differences between the results of NINA and SIS can be attributed to the different time period (since the flux of helium is known to change significantly over the months) and to the fact that the SIS flux include also ^{3}He.

3.1. Solar Modulation of Cosmic Rays

In Figure 2 we show a comparison of the data before Solar Maximum (1999) and during Solar Maximum (2000). Helium flux, which in this range is also due to the anomalous component[8] is greatly reduced by the increased solar activity.

---

[a]The McIlwain parameter L represents, at a first approximation, the value (expressed in Earth radii) at which the magnetic field line passing through the point considered intersects the geomagnetic equator. In the case of low Earth orbits (such as Mir) values close to L =1 are in proximity to the equator and increase at higher latitudes. For a detailed definition see [7].

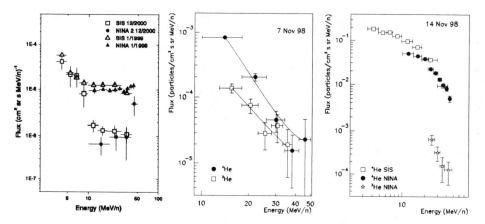

Fig. 2. *Left Panel:* ^4He measurements taken before (1999) and at solar maximum (2000), inside (NINA-1 and -2) and outside (ACE - SIS) the magnetosphere. Note the reduction of anomalous (10-50 MeV) helium due to the increased solar activity. *Centre and Right Panels:* ^4He and ^3He differential spectrum observed in the 1998 solar events of 7/11 (Left) and 14/11 (Right). Note the difference in fluence between the two events and the different He isotopic ratio.

4. Solar Energetic Particles

Solar Energetic Particle (SEP) events consist in a rapid particle flux increase (up to 4 orders of magnitude) emitted from active regions on the solar surface. It is currently believed that magnetic line recombination due to thermal motion of magnetic field frozen with the solar plasma can produce energy necessary for particle acceleration (typical released energies are $E = 10^{22}$ $10^{25} J$ over a region of length $\simeq 3 \times 10^7 \, m$ and height $\simeq 2 \times 10^7 \, m$). These energies are released in a few seconds during solar flares, which consist an intense brightening (in visible and X-ray) of the solar surface, usually close to sunspots. It was therefore natural to believe that all particle acceleration took place at flare site. However recent studies are now moving toward a (blurred) division in two categories[11]:

- *Flare accelerated events*, where particles are accelerated at the flare site; they are usually small, impulsive, of low intensity and present an high $^3He/^4He$ ratio and heavy nuclei enhancement.
- *Coronal Mass Ejections*, where particle acceleration occurs at the shock front: they are of gradual nature and more intense, with little or no isotopic or nuclear increase over solar abundances.

This division is however somewhat artificial; there is agreement over the fact that there is no "typical" case and each event has its own peculiarities. This is also due to the fact that in addition to the different production and acceleration processes, there are also propagation effects in the interplanetary medium. In case of the smaller events, the kinetic energy of accelerated particles is small compared to the solar wind kinetic and magnetic energy density: particles therefore gyrate along the archimedean spiral of the interplanetary magnetic field as they leave the sun.

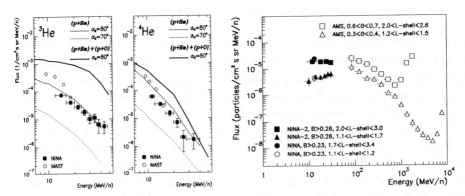

Fig. 3. Left and Centre panels: South Atlantic Anomaly ^{3}He and ^{4}He data measured with NINA-2 and MAST ($L = 1.181.22$, $B < 0.22G$). Right panel: NINA-1, -2 and AMS data of the secondary proton component of cosmic rays at low latitude. It is possible to see the effect of the geomagnetic cutoff in the AMS points taken at L=2.0 L=2.6.

These events can be observed only if the probe (or Earth itself) is located along the magnetic field line corresponding to the emission site. In case of CMEs the kinetic energy of the emitted particles is intense enough to distort the magnetic field lines as the shock travels in the interplanetary space. CMEs can thus be observed over a much wider longitude range; the intensity vs time profile of the same event is very different according to the observation point: it is sharper in case of direct magnetic connection to the emission region and more gradual toward the boundaries.

NINA-1 and 2 observed more than 30 Solar Particle Events[b], analyzing the particle flux, differential spectrum and nuclear and isotopic composition. The event of 1998 November 14 was among the most powerful, with counting rate increase of almost 3 orders of magnitude with respect to solar quiet periods. For the other events we registered increases of one or two orders of magnitude on average. The spectral index γ for ^4He and the isotopic composition are different between these two events, ranging from 1.8 in the 1998 November 14 event to 6.8 in the 1998 November 8 event (Figure 2). The ^{3}He spectrum, also fitted by a power-law, is slightly harder ($\gamma = 2.5 \pm 0.6$) than that of ^{4}He ($\gamma = 3.7 \pm 0.3$) resulting in an ^{3}He/^{4}He ratio increase with energy[5].

5. Trapped and secondary component of Cosmic Rays

Earth's magnetosphere is the only region where is possible to study with *in situ* measurements collisionless shocks and the phenomena related to the interaction between solar wind and the geomagnetic field. Knowledge gathered directly in this regions can be applied not only to an understanding of the heliopause but also to modelling all effects where magnetohydrodynamics plays an important role: from

---

[b]The launch itself of NINA-2 was performed the day following the 14th July 2001, when particle flux was 10^4 times higher than solar quiet period.

solar energetic particle acceleration to supernova shocks to interaction between particles and neutron star magnetic fields.

One of the most notable effect of the interaction of cosmic rays with the geomagnetic field is the phenomenon of cosmic rays trapping in radiation belts. Since the magnetic field exerts no work on particles, the trapped region is unaccessible to primary cosmic rays: trapped protons come mostly from neutron decay while electrons are trapped by temporal variations of the geomagnetic field.

Another mechanism which can produce trapping or quasi-trapping is secondary particle production in the atmosphere. This is the case - for instance - of Helium: Figure 3 shows the ^{3}He and ^{4}He NINA and MAST measurements in the South Atlantic Anomaly (SAA: $L = 1.181.22$, $B < 0.22G$). The spectral indexes are $\gamma(^3He)$ = 2.30 ± 0.08 in [10–50 MeV] and $\gamma(^4He) = 3.4 \pm 0.2$ in [10–40 MeV]. The solid lines are calculations which hypothesize the interaction of trapped protons with Helium and Oxygen of Earth's atmosphere[9]. It is possible to see that the inclusion of Helium in the secondary particle production overestimates the ^{3}He production. Interaction with the atmosphere is a process which has an important role also in the production of secondary particles, usually found below the expected cutoff[c]. This phenomenon was originally described, for instance, in [12]; recent measurement of the secondary component by the AMS[13] detector has provided new information on this component confirming its secondary nature also by calculations[14]. A detailed understanding of this component is not only important for the comprehension of the interaction between the geomagnetic field and cosmic rays but also for background estimation of future gamma ray experiments such as Agile and Glast. In Figure 3 (right) are shown the measurements of NINA and AMS in different orbital regions. It is possible to see how the differential spectrum varies significantly as a function of energy according to cross section in the atmosphere and propagation effects. The two data set are in qualitative agreement; from the comparison of NINA and NINA-2 data (the former taken before solar maximum and the latter at solar maximum) it is possible to see how the effect of solar modulation is small, giving credit to the hypothesis of a high energy (and thus unmodulated) primary component which interacts with the top layers of the atmosphere.

6. Measurements on board Mir Space Station

These measurements have been carried out with two silicon detector telescopes, Sileye-1 and Sileye-2. These devices have been designed and built to study the cosmic ray radiation inside spacecraft and to relate the different nuclear composition with external particle flux. In addition to the importance to cosmic ray physics these measurements are necessary for a detailed assessment of the absorbed and

---

[c]The geomagnetic field deflects galactic and solar cosmic rays: the *geomagnetic cutoff* G represents the impulse below which an orthogonally incident particles of charge z cannot reach the Earth. This value is given at first approximation by $G = 14.9\ z\ cos^4\lambda (Gev/c)$, where λ represents the geomagnetic latitude.

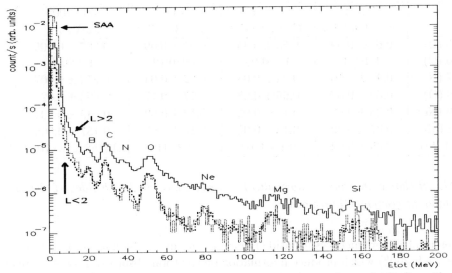

Fig. 4. *High energy nuclear abundances inside Mir: Continuous line: galactic ($L > 2$) component; Dotted line: SAA component ($L < 2$, $B < 0.25G$); Dashed line: remaining region ($L < 2$, $B \geq 0.25G$). The SAA region has a higher proton flux due to trapped particles but $Z > 5$ particles are equally abundant to the $L < 2$ region due to the equivalent cutoff.*

equivalent doses of the astronauts in conditions of solar quiet and active days. Another item is the investigation of the Light Flash phenomenon, which consists of the observation of visual phenomena triggered by cosmic rays by means of a yet unidentified mechanism[19].

Sileye-2 consists in a silicon detector telescope, housed in an aluminium box (total mass 5.5 kg), coupled to an helmet with an eye mask. The particle detector telescope is made of six silicon active wafers identical to those used for the construction of NINA-1 and -2, resulting in three double wafer planes and 96 independent readout channels. The distance between planes is 15 mm, resulting in a geometrical factor of 85 $cm^2\ sr$.

In space, astronauts and equipment may be subject - according to the orbit - to intense fluxes of radiation. All manned flights took and take place in Low Earth Orbit (LEO), at altitudes varying between 200 and 500 km, with the only notable exception of the Apollo program, which not only took man on the Moon but also outside Earth's geomagnetic shielding. Most LEO missions (such as Mir, the International Space Station and Shuttle) have a low inclination (51.6°) to avoid high latitude areas, where the lower geomagnetic cutoff results in an higher cosmic ray flux. Cosmic ray and radiation measurements inside spacecraft have been the subject of intense investigation throughout the course of space exploration. These studies are particularly difficult since they need to take into account the orbital dependence of cosmic ray flux and its propagation inside the varying absorber thicknesses of the spacecraft. Investigations therefore include dosimetric measurement (often with

Z	$L > 2$ $(C > 0.6\,GV)$	$L \leq 2$ $(C > 3.9\,GV)$	SAA $(C > 3.9\,GV)$	600-1000 MeV/n [15]
5 (B)	0.63 ± 0.09	0.53 ± 0.35	0.55 ± 0.09	0.307 ± 0.005
6 (C)	1 ± 0.1	1 ± 0.06	1 ± 0.12	1 ± 0.02
7 (N)	0.41 ± 0.06	0.34 ± 0.08	0.22 ± 0.04	0.274 ± 0.007
8 (O)	0.65 ± 0.07	0.66 ± 0.08	0.77 ± 0.17	0.93 ± 0.02
10 (Ne)	0.33 ± 0.06	0.13 ± 0.02	0.12 ± 0.02	0.149 ± 0.004
12 (Mg)	0.07 ± 0.02	0.13 ± 0.02	0.12 ± 0.02	0.187 ± 0.005
14 (Si)	0.05 ± 0.02	0.1 ± 0.02	0.1 ± 0.02	$.13158 \pm 3 \cdot 10^{-5}$

Table 1: *Relative abundances normalized to carbon measured with Sileye-2 detector in different geomagnetic regions.*

different absorber thickness or inside human phantoms to simulate propagation inside the body[16]) which produce information on the total absorbed dose. A number of more sophisticated detectors are also used to monitor in real time long and short term variations in space[17].

In addition to orbit dependent dose differences it is possible to study the differences of nuclear composition in different points of the orbit[18] to show latitude dependent effects. The orbit of Mir was divided in three regions: Galactic Cosmic Ray region (GCR, $L > 2$), South Atlantic Anomaly (SAA: $L < 2$ and geomagnetic field $B < 0.25G$), and the remaining region ($L < 2$, $B \geq 0.25$ G). In a given point of the orbit, the geomagnetic cutoff C determines the minimum energy for primary cosmic rays to reach Mir and be detected. This value is valid for particles orthogonal to the local field line and outside Mir. In addition, particle energy inside the station can be modified by the interposed material of the station and the presence of nuclear interactions, so it should be used only as a reference. Only particles with energy above 90 MeV/n (of which at least 20 Mev/n are lost in the Al of the station) are being considered. Naturally the 3mm Al thickness assumed only represents a lower value, since different amounts of the station and the equipment contained can be interposed between the detector and the local field line along which the particles come. At $L = 2$, $C = 3.9\,GV$ while at high latitude (L=4.4) $C = 0.8\,GV$. These two values represent the minimum cutoff for a given region; they correspond to a minimum kinetic energy (for particle with mass/charge ratio of 2) of $\approx 150\,MeV/n$ ($C = 0.8\,GV$) and $\approx 1600\,MeV/n$ ($C = 3.9\,GV$). At these energies particles lose only a small fraction of their kinetic energy crossing the hull of the station: again, using 3 mm of Al as a reference, we find that, for instance, a 150 (1600) MeV proton loses 3.4 (1.3) MeV to enter the station. In case of other nuclei, the values are similar: if we consider a carbon nucleus, we have 4.8 (1.2) MeV lost for 150 (1600) MeV/n. The particle distributions of the three regions are shown in Figure 4. The continuous line shows the galactic component which has an higher flux due to the lower geomagnetic cutoff. This allows particles with lower energy to reach

Mir resulting in a higher particle count. The wider energy range implies a larger energy release range, resulting in the peaks to be less sharply defined. In this range, proton and helium flux is lower than that measured in the SAA (dotted line) where the trapped component is dominant if compared to galactic and $L < 2$ abundances. Indeed the $L < 2$ curve (dashed) has a lower $Z \leq 2$ flux if compared to SAA but an equal $Z \geq 5$ flux, since in both cases the component selected at this energy is the same. From these distributions it is possible to reconstruct relative abundances for the different nuclear species (shown in Table 1, compared with relative cosmic ray abundances at 1AU[15] measured in the energy range of $\approx 1 GeV$. It is possible to see how, especially for the $L < 2$ regions, notwithstanding the bulk of the Mir, the data are in general in agreement. There are however several important differences: 1. An overabundance of B in respect to C. It is roughly twice the 1AU value in all three regions. 2. A higher amount of N in the $L > 2$ region compared to the other regions and 1AU data, again due to an larger production of secondary N at lower energies. 3. A lower amount of Oxygen nuclei in $L > 2$ and $L < 2$ regions (SAA value is in agreement within errors with 1 AU data). In all cases the most probable cause are hadronic interactions in the hull of the space station, resulting in secondary particle production of B and N. The underabundance of O in respect to C is due to its higher cross-section: Oxygen could be considered as composed of four alpha particles (He nuclei) and Carbon of three. Thus, if we assume the ratio of the cross sections to be equal of the ratio of the nucleons of Carbon and Oxygen ($12/16 = 0.75$) and we multiply by the original flux of 0.93 we obtain an abundance of 0.7.

7. Current work

Work in the field of cosmic ray physics from space is continuing in parallel with the construction and launch of two larger experiments: Pamela and Sileye-3/Alteino.

Pamela is a magnetic spectrometer[20] with the primary goal of study the antimatter component of cosmic rays. Secondary objectives are the continuation of NINA observations of the low energy component in the period going toward solar minimum and the observation of the positron spectrum in solar energetic particles. Launch is foreseen by the end of 2002; the expected lifetime is of three years.

Sileye-3/Alteino is to be placed on the International Space Station with the taxi flight of the italian astronaut Roberto Vittori in April 2002. Its goal is to continue to obervations of the previous Sileye experiments monitoring the nuclear abundances and measuring the absorbed and equivalent dose inside ISS. The studies on Light Flashes will continue with the use of an Electroencephalograph during the permanence of Vittori in space, while cosmic ray monitoring will continue for the next years.

Data coming from the two experiments will be correlated among themselves and with other satellites in order to study propagation effects in the heliosphere and within Earth magnetosphere.

Acknowledgments

I would like to thank Dr. V. Mikhailov and Dr. R. Sparvoli for the useful discussions and the help in the preparation of this work. A special word of thanks goes to Eng. M. Crisconio for the invaluable help during NINA-2/MITA mission planning and data taking.

References

1. Golden, R. L., et al. , *ApJ* , **436**, 769, 1994; Golden, R. L., et al., *ApJ* , **457**, L103, 1996; Boezio, M., et al., *ApJ* , **487**, 415, 1997.
2. Casolino, M., et al., *Proc. ICRC*, 2314, 2001.
3. Bakaldin, A., et al. 1997, *Astrop. Phys.*, **8**, 109.
4. Bidoli, V., et al., 1999,*NIM A*, **424**, 414; Bidoli, V., et al., *ApJ supp*, **132**, 365, 2001.
5. Sparvoli, R., et al., *Proc. ICRC*, 3104, 2001.
6. P. K. F. Grieder, *Cosmic Rays at Earth*, Elsevier, 2001.
7. McIlwain, C. E., *J. Geophys. Res.* 66 3681, 1961.
8. Simpson, J. A., *Sp. Sci Rev.*, **83**, 7, 1998; Cummings, A. C., et al, Stone, E. C., *Sp. Sci Rev.*, **83**, 51, 1998.
9. Selesnick, R. S., Mewaldt, R. A., *JGR*, 101, 19745, 1996.
10. Mason, G. M., Mazur, J. E. & Dwyer, J. R., *ApJ*, **525**, L133, 1999.
11. Reames, D., Sp. Sci. Rev. 90, 413, 1999.
12. Treiman, Phys. Rev. 91, 957, 1953; Perola G.C., Scarsi L., *Nuovo Cimento*, **44**, 4, 718, 1966; Bland, *Space Res.* , **5**, 618, 1965.
13. Alcazar, J., at al., *Phys. Lett. B*, **472**, 215, 2000.
14. Derome, L. et al., astro-ph/0103474v2, July 2001; Lipari, P. *Astrop. Phys.*, 367, 2002.
15. Simpson, J. A., *Ann. Rev. Nucl. Part. Sci.*, **33**, 323, 1983.
16. Badhwar, D., et al, *Rad. Res.*, **149**, 209, 1998.
17. Reitz G., et al. *Radiat. Meas.*, **26 6**, 679, 1996; Sakaguchi T., et al., **NIM A** , **437**, 75, 1999.
18. Casolino, M. et al., *Proc. ICRC*, 4011, 2001; Bidoli, V., et al., *J. Phys. G*, **27** 2051, 2001.
19. Casolino, M., *Proc. of Aquila School on Space Science*, 2002.
20. Spillantini, P., et al., *Proc. ICRC*, 2215.

LOW ENERGY ELECTRON AND POSITRON SPECTRA IN THE EARTH ORBIT MEASURED BY MARIA-2 INSTRUMENT

V. V. MIKHAILOV

Moscow Engineering and Physics Institute,
31, Kashirskoe shosse Moscow, 115409, Russia

Results of study of secondary electrons and positrons in the energy range 15–150 MeV in the earth space-orbit are reviewed.

Keywords: Electron; Positron; Geomagnetic field; Cosmic Rays Albedo .

1. Introduction

Cosmic ray interactions with residual atmosphere produce secondary electrons, positrons, nucleons and photons. These secondaries make their trajectories in the Earth's magnetic field. If the particle's rigidity is less than the geomagnetic cut-off, such particles can be kept by the field. This phenomenon has been known since the first balloon and rocket experiments. For example, the work of Gangles at al. (Ref. 1) contains results of particle measurements up to 150 km made with Geiger counters and telescopes. The count rate of particles versus altitude on both ascending and descending parts of the rocket trajectory showed that a substantially flat plateau intensity exists above about 55 km. For the authors it was obvious that "secondaries emerging from the atmosphere undoubtedly contribute in ... the intensity above the atmosphere". However there were no quantitative theories which took into account the magnetic field, and there was speculation that some part of fluxes above the atmosphere could have a cosmic origin. Another indicative information came from high energy photon measurements aboard satellites. The angular distribution of photons near the equatorial region at L-shell ~ 1.1, as measured by the instrument on board OSO-3 satellite [2], has a peak toward the horizon which is about one order of magnitude more than intensity in the nadir direction. Similar results were obtained with SAS-2 satellite [3]. From these observation it is clear that horizontal electromagnetic showers play an important role to the formation of the radiation environment around the Earth.

Numerous satellite experiments were carried out in the years 1960–1970, aimed at studying the secondary electron component at high altitudes. Some of these are listed in figure 1. To explain all the experimental results that such experiments collected it was suggested by Grigorov [4] that the interaction of primary cosmic rays with the residual atmosphere might create electron fluxes with energies of several

authors	Year, Spacecraft	Altitude, km	Energy, MeV
Basilova, Grigorov et al	1966, Proton-3	150-600	Ep>30, Ee>5
Basilova, Grigorov et al	1968, Proton-4	225-495	Ep>30, Ee>5
Kurnosova et al	1968, Cosmos 225	257-530	Ee>125
Basilova, et al	1972, Cosmos 490	210-305	Ee>80
Kurnosova et al	1973, Cosmos 555	216-253	Ep>400, Ee>10
Galper, et al	1979, Salut 6	300-350	Ee=30-350
Galper, et al	1982, Salut 7	300-350	Ee=30-350
Galper, et al	1981, Cosmos 1300	825-906	Ee=20-350
Galper, et al	1987, MIR*	350-400	Ee=20-150
Galper et al	1991, Gamma 1	400	Ee=50-40000
Abramenko et al	1991, Cosmos 1870	450	Ee=8-36

Fig. 1. List of experiments which measured secondary electrons in the proximity of the radiation belts.

hundred MeV, trapped by the geomagnetic field. In Grigorov's work a mathematical analysis of the likelihood and features of such a radiation belt was done, and it was shown that the intensity of the flux of secondary electrons may be of the order of the intensity of primary cosmic rays, and is not strongly dependent on the altitude. Due to the geomagnetic field, secondary particles may drift many times around the Earth if the drift density is quite small. Albedo particles cannot make more than one bounce if mirror points are deep in atmosphere. There are also quasitrapped particles making many bounces but only one drift around the Earth. According to Grigorov's model, the number of electrons produced is proportional to the cosmic ray flux I_{CR} and to the probability of their interaction in residual atmosphere, which is itself proportional to the density. On the other hand, if the life-time T of secondary particles is determined by their energy losses, which are mainly ionization and radiation, T is proportional to $1/\rho$ (this assumption is valid if $\rho > 10^{-15}$ g/cm^2). As a result, the flux in the Earth magnetic field is independent from the density (or, the same, from altitude).

The secondary flux can be described by the following equation[5]:

$$J(>E) = I_{CR}(l)K(l,a)\int X(E,E')n(E',l,a)dE'/L , \qquad (1)$$

where $I_{CR}(l)$ is the cosmic ray flux at the latitude l, and L is the range of the cosmic ray protons for the inelastic interaction. $K(l,a)$ is a coefficient that takes into account the probability of the production of secondary electrons with pitch-angle a at the latitude l. X is the amount of residual atmosphere in which electrons with energy E' lose energy $E' - E$, $n(E',l,a)$ gives the source function (number of the electrons in the energy range $E' \div E' + dE'$ and in given direction produced in one interaction at the latitude l). The equation 1 was used to estimate the secondary electron flux, assuming that the production and decay of charged pions gives the major contribution to the source function n. The secondary electron flux calculated according to 1 is independent from an atmospheric altitude of several hundreds kilometers, and has a weak pitch-angle dependence. As result of that fluxes trapped, quasitrapped and albedo particles are closed. [4,5]

The serious limitation of this model consists in the fact that it makes use of only one possible process (π-μ-e process) and neglects secondary propagation of particles, which includes not only energy losses but also particle production in electromagnetic showers. The secondary propagation is, instead, important especially for the albedo flux (particles with small pitch-angles) calculation. Detailed calculations of interactions of primary cosmic rays with the Earth's atmosphere, which take into account all main processes, were performed for vertical direction in the energy range from several MeV to tens of GeV [6,7]. It was shown that the cascade development determines the energy spectra of particles at low energy (E<100 MeV). A three dimensional model was considered for gamma-rays with energy more than 30 MeV [8]. It is evident from the results of this work that electromagnetic showers are important even at the top of atmosphere to explain the angular distribution of gamma rays. Calculations in Ref. 6-8 were done without considering the influence of the Earth' magnetic field. This prevented the authors from adequately analyze the escape of electrons from the atmosphere into the Earth-orbital space.

Since 1980 there were a set of experiments done by the MEPhI group, which were mainly devoted to the investigation of the electron and positrons fluxes in SAA. One can find a review of the results in the paper Ref. 9, and some of these experiments also are listed in the table of figure 1. Direct measurements of the charge composition of the electron component by MARIA-2 instrument aboard MIR station provided new data to improve theoretical models of electrons environments in the near Earth orbit. This paper presents a review of low energy electron and positron measurements, and their interpretations.

2. Instrument

The MARIA-2 layout is sketched in figure 2. The C1, C2, and C3 hodoscope scintillation detectors measure the angle of a charged particle deflection in magnetic field generated by a permanent magnet M. The magnet is surrounded by an anticoincidence scintillation counter AC. Detectors C1 and C3 constitute a time-of-flight system which separates downward and upward moving particles. This system measures the time of flight with resolution of about 1.5 ns. In this way it measures the particle velocities. Knowing both the particle time of flight and magnetic deflection angle (in magnitude and sign), it is possible to select electrons and positrons from protons. The energy range of instrument is 15–150 MeV for electrons and positrons. The instrument is also capable of measuring fluxes of 35–100 MeV protons. The

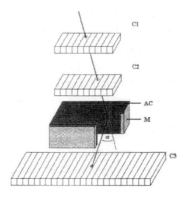

Fig. 2. A schematic layout of MARIA-2 instrument.

C1 and C2 hodoscope detectors allow measurements of the particle incident angle with an accuracy of about 2^0 on X coordinate and of about 4^0 on Y.

The data acquisition is performed in orbit in real time. The MARIA-2 acquisition system, based on a microprocessor, analyzes each recorded event, supervises all detectors and records data in the memory. The instrument characteristics were determined by Monte-Carlo calculations using parameterizations referred in Ref. 11 and were then verified on beam test. A more detailed description of the instrument is given in Ref. 10.

3. Experimental results

Measurements on board MIR station have been made periodically since 1988. MARIA-2 steadily monitored the particles fluxes present in the vicinity of the Earth. Owing to the MIR station orbit (about 400 km altitude and 51.7 inclination) it was possible to detect secondary electrons and positrons generated in residual atmosphere. In the South Atlantic Anomaly (SAA) region the lowering of altitude of the inner boundary of radiation belt favors the measurement of particles trapped by the geomagnetic field.

The electron and positron energy spectra of albedo fluxes were measured throughout the MIR orbital path. Previous experiments on board satellites, like PROTON-3, PROTON-4, COSMOS-490, COSMOS-555, InterCOSMOS-1300, orbital stations SALUT-6 and SALUT-7 showed that fluxes of high energy secondary electrons (sum of electrons and positrons) have a complex spatial structure and depend on longitude, latitude, pitch-angle. Figure 3 shows the differential energy spectrum of the sum of electrons and positrons measured near the equator (geomagnetic rigidity R > 10 GV) by different experiments. MARIA-2 results on figure 3 show good agreement with the other data. Recent measurements [18] done with AMS instrument are also shown. One can see that outside the SAA region the spectrum turns out to fit rather well with a $E^{-\gamma}$ law with the index γ about 2 in the wide energy range from some MeV to several GeV. The solid line on this figure presents results of Grig-

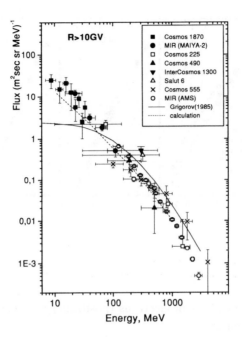

Fig. 3. Differential

orov calculations[5], the dotted line is discussed below. It is evident that Grigorov's $\pi - \mu - e$ decay model overestimates the flux of secondary electrons at high energies. Since the calculation in Ref. 5. was done for an isotropic cosmic ray distribution ($K(l,a)=1$ in Eq. (1)), this result is not surprising. There is also difference from experiments at low energy, where the model spectrum [5] is practically flat whereas the experimental one is not.

Figure 4 shows the energy spectra of electrons and positrons measured beyond the SAA for different L-shells. One can see that outside the SAA region the spectra of albedo electrons and albedo positrons are nearly identical at energy below 100 MeV. This charge composition can not be explained solely by the process of $\pi - \mu$-e decay. As it is known from calculations, the ratio $\pi+$ /π- for pions generated by primary cosmic ray is about 1.4-2.0 in this energy range [6,7].

4. Calculations

To improve the theoretical model new calculations of albedo fluxes have been undertaken. A three dimensional study has been carried out by means of a Monte Carlo method. In contrast with previous calculations the modelling of processes of interaction and propagation gives the possibility to take into account the angular distribution of particles, kinematics of the reactions, energy losses and secondary production. Let us take points with invariant geomagnetic latitude Θ, longitude Φ

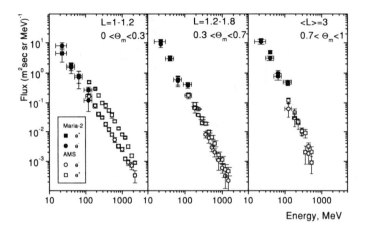

Fig. 4. Differential energy spectra of electrons and positrons measured by MARIA-2[12,14] and AMS[18]. L-shell intervals are shown for MARIA-2 and invariant geomagnetic latitude Θ for AMS.

and altitude h. The numbers of electrons (positrons) with energy E produced in volume dV in direction w is:

$$N(\Phi,\Theta,E,h,w)dEdVdw = \int I_{CR}(\Phi,\Theta,E',\Omega,h)W(h,E,E',w,\Omega)dE'dEdVd\Omega dw. \quad (2)$$

where I_{CR} is intensity of nuclei, W is number of electron (positron) production with energy E in direction w by primary particles going in direction Ω with energy E'. There is conservation of Θ during particle propagation if the particle's energy much less than the geomagnetic rigidity. With this assumption the averaged flux over longitude for chosen point of space ξ is given by:

$$F(\Theta,E,\xi,w) = \int n(\Theta,E,\xi,w,E',w',\xi')N(\Theta,E',h',w')dE'dw'dV'. \quad (3)$$

where n is "probability" to observe in point ξ secondary particle produced in direction w' in ξ' with the energy E'. Integration is done along magnetic tube. Due to limited computer speed calculations could be performed only albedo component.

The intensities of secondary particles were calculated by the Monte Carlo method under the following assumptions: the intensity of high energy cosmic ray is isotropic at the boundary of the atmosphere in the upper hemisphere; the intensity at the level x of the residual atmosphere is determined by $I(E,\theta,\varphi,x) = I(E)exp(-x(\theta)/l)$, where θ is zenith angle, φ is azimuth angle, l is attenuation length (120 g/cm^2 for protons), E is the energy of cosmic particles $E > R(\theta,\varphi)$, where $R(\theta,\varphi)$ is the geomagnetic rigidity. To calculate the total intensity the summation was done for primary and secondary protons and α particles. The energy spectrum $I(E)$ is dependent on solar activity. The geomagnetic rigidity was calculated for a dipole magnetic field. The density of the atmosphere is assumed to be exponentially dependent on altitude; the composition of the atmosphere is independent of altitude

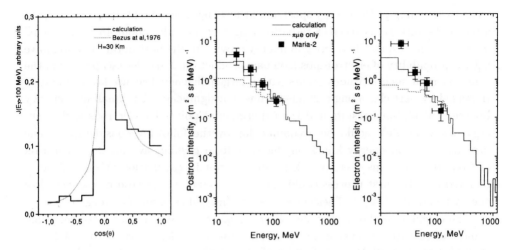

Fig. 5. Left: Measured in a balloon experiment and calculated angular distributions of gamma rays with energy more 100 MeV [13]. Center and Right: Differential energy spectra of electrons and positrons measured near geomagnetic equator[12] (L-shell < 1.2, B > 0.24 Gs). Solid line is result of calculations, dashed line presents results only for charge pion decay source.

above 200 g/cm^2. Approximation formulas were used to describe the energy and angular distributions of pions generated in interaction of cosmic rays with air nuclei. The angular distribution of particles was taken into account for the simulation of pions and muon decays. The simulations of trajectories of electrons and positrons were performed by numerical integration of the equations of motion in the geomagnetic field. The processes of ionization, productions of knocked-on electrons, bremsstrahlung, multiple Coulomb scattering and electron-positron pair production were taken into account for simulation of the electron and positron propagation in the atmosphere. The details are described in the works [12,13].

5. Result of simulation

Figure 5(left) shows the calculated gamma ray angular distribution at an altitude of 30 km compared with experimental data. Also there are calculated vertical fluxes of muons and positrons for different altitudes and geomagnetic latitudes [12,13]. These calculations and comparisons show that the source function used in this work is in agreement with the other data. Figure 5 (center and right) shows the differential energy spectra of albedo electrons and positrons (pitch-angle < 50^0), for L-shell=1.1 (R=12,5 GV). Experimental points are the data of MARIA-2 instrument on board MIR station. The calculated spectrum of the combined flux of electrons and positrons is approximated by a power law dependence with index of -1.9 for energy more than 50 MeV. Dashed histogram in this figure shows the fraction of electrons (positrons) produced directly by the decay of charged pions generated in cosmic ray interaction with the Earth's atmosphere. It seems from the figure that the role of secondary gamma rays is very significant at energies less than 50 MeV. It leads to softening of the energy spectrum at low energies. The atmospheric layer in a horizontal direction exceeds a radiation length at altitudes about 45 km (1 g/cm^2 in vertical direction). Therefore gamma ray moving in the horizontal direction have a great probability of electron-positron pair production. Particles escaping from the atmosphere at such altitudes occurs with a large efficiency, because layer of matter traversed by particles along a magnetic line is insignificant. The small discrepancy between calculations results and the experimental points for electrons with a energy below 30 MeV could be accounted for by the following consideration. The production of low-energy knock-on electrons from cosmic ray and the generation of Compton electrons was not taken into account in calculation. Meanwhile it is believed that the first process could determine secondary electron flux at very low energy below 10 MeV [15]. These processes are insignificant for positrons.

The calculation shows that the observed spectrum of electrons (positrons) becomes softer with increasing latitude. The spectral index is equal to -2.4 for L-shell=1.8 at E > 50 MeV. Softening of the spectrum is connected with changes in the conditions of particle leakage from the atmosphere due to the increasing of the magnetic line inclination. The angle between the directions of primary and secondary particles should be large at high latitudes in order to let the secondaries escape from atmosphere. At high energy secondary particles go mainly in primary particle direction. Since the angular distribution becomes more isotropic with lowering of energy, low energy particles still escape even in zenith direction.

6. Discussion

It is obvious from figure 3, where differential energy spectrum of leptons are shown, that a model which takes into account secondary propagation and angular distribution (dashed line on this figure) adequately reproduces the experimental data. The calculations are also in good agreement with new experimental results.

Figure 4 shows electrons and positrons spectra separately measured by AMS and MARIA-2 instruments for different L-shells. Data of both experiments could be fitted by one law except for one point at high energy in MARIA-2 spectrum. This MARIA-2 point could be affected by proton background. No subtraction of background was done in data. At equatorial region background was found to be negligible [13], meanwhile at middle latitudes the proton flux increases and could contribute to the high energy point of the lepton spectrum due to a proton scattering inside the instrument.

One can note that the electron-positron ratio at equatorial regions (L-shell=1.2) differs between AMS and MARIA-2. It is more clear from figure 6. This figure shows the electron-positron ratio versus energy. It was believed that large value of this ratio measured by AMS experiment is explained by the East-West effect [16,17]. Due to the influence of the geomagnetic field, escaping of produced positrons from atmosphere is more preferable than electrons in equatorial regions. It is obvious that with energy decreasing this effect becomes small because low energy secondary particles have

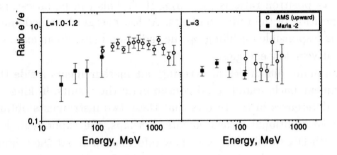

Fig. 6. Electron to positron ratio versus energy in two different intervals of geomagnetic latitudes

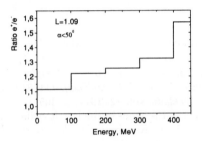

Fig. 7. Calculated electron to positron ratio versus energy for albedo particles for L-shell=1.09

practically isotropic angular distribution. So one could expect that this ratio is a function of energy. At first, it is increasing with energy and then should decrease at very high energies. This tendency is presented in figure 6. However detailed calculations of this behavior are absent in the 10–100 MeV energy range. East-West effect is less at high latitudes and, as it should be, the agreement between experiments is better (see figure 4 and 6).

There is no East-West effect for albedo particles that have small gyroradius and do not drift in geomagnetic field. In this case the ratio observed is connected only with secondary production cross sections and secondaries propagations. The calculated ratio is shown in figure 7. There is increasing with energy from about 1 at E=10 MeV to about 1.5 at 400 MeV.

As for the measurements in SAA, results of MARIA-2 experiment are presented in papers Ref. 9. It was found that there is excess of electrons over the cosmic ray albedo electrons in inner radiation belt. The trapped positron flux is similar to the albedo flux within statistical fluctuations. On the other hand preliminary results of AMS experiment [E. Fiandrini, private communication] show that trapped electron and positron fluxes have the same charge ratio as quasitrapped secondary flux. In other words there is excess of positrons over electrons in SAA and the electron to positron ratio is less than unity. There are also estimations of this ratio at very low energy in keV interval, made during particle precipitations [19]. It was found that the electron to positron ratio is $(4 \pm 3) \times 10^{10}$. This can be an estimation of the charge composition of the radiation belts. At low energy the albedo neutron decay mechanism is responsible for filling up the radiation belts. So an excess of electrons is expected at very low energies.

There are many difficulties in carrying out measurements inside the radiation belts. Instrument background could contaminate the results. Taking into account the existing differences in the results that these two instruments obtain, it is necessary to make additional measurements. Perhaps, the experiment PAMELA on board RESURS-DK satellite can give new information about the origin of trapped leptons since its energy threshold is lower than AMS and the background separation is better than for MARIA-2.

References

1. A. V. Gangnes, J. F. Jenkins and J. A. Van Allen, *Phys. Rev.* 75, 1, 57 (1949).
2. W. L. Kraushaar, G. W. Clark, G. P. Garmiree et al., *ApJ* 177, 341 (1972).
3. D. J. Thompson, G. A. Simpson, M. E. Ozel, *J. Geophys. Res.* 86, 1265 (1981).
4. N. L. Grigorov, *Akademiia Nauk SSSR, Doklady*, 234, 810 (1977) (in Russian). Translation :*Soviet Physics - Doklady*, vol. 22, 1977, p. 305.
5. N. L. Grigorov, "High energy Electrons in the Earth vicinity", Nauka, Moscow, 1985 (in Russian). N. L. Grigorov, L. V. Kurnosova, L. A. Razorenov, M. I. Fradkin, Preprint of Lebedev Phys. Inst., Moscow, n. 219, 1987.
6. K. P. Beuermann, *J. Geophys. Res.*, 76, 4291 (1971).
7. R. R. Daniel, S. A. Stephens, *Rev. Geophys.*, 12, 233 (1974).
8. D. J. Thompson, *J. Geophys. Res.*, 79, 929 (1974).
9. A. M. Galper et al., "Radiation belts: models and standards", Geophysical monograph, 1997.
10. S. A. Voronov, A. M. Galper, S. V. Koldashov et al., *Prib. i Tekhn. Eksp.*, 2, 59 (1991)(in Russian).
11. A. N. Kalinovskii, N. V. Mokhov, Yu. P. Nikitin, "Passage of high energy particles through matter", (Energoatomizdat, Moscow, 1985) (in Russian). Translation: AIP edit, New-York, 1989.
12. S. A. Voronov, S. V. Koldashov, V. V. Mikhailov, Preprint of Moscow Eng. Phys. Inst., Moscow, 16-92, 1992 (in Russian). S. A. Voronov, S. V. Koldashov, V. V. Mikhailov, *Cosmic Res.* 33, 3, 302 (1995).
13. V. V. Mikhailov, PhD thesis, Moscow Eng. Phys. Inst., Moscow, 1993, (in Russian).
14. S. A. Voronov, A. M. Galper, S. V. Koldashov, L. B. Maslennikov et al. *Kosmicheskie issledovania* 21, 4, 567 (1991), (in Russian).
15. A. A. Gusev, G. I. Pugacheva, *Geomagnetism and Aeromomia*, 6, 912, (1982) (in Russian).
16. L. Derome, M. Buenerd, Yong Liu, *Phys. Lett.* B. 521, 139 (2001).
17. P. Lipari *Astroparticle Physics*, 16, 295 (2002).
18. J. D. Alcaraz et al., *Phys. Lett.* B, 484, 10 (2000).
19. L. E. Peterson, *J.Geophys. Res.* , 69, 15, 3141 (1964).

THE TRAPPED ANOMALOUS COMPONENT OF THE COSMIC RAYS : THE SHORT OVERVIEW OF EXPERIMENTS

M.I. PANASYUK*

*Skobeltsyn Institute of Nuclear Physics of Moscow State University
119899, Vorob'ovy Gory, Moscow, Russia*

The experimental data on the chemical composition of galactic cosmic rays (GCR) are analysed in the energy range between $\approx$10 MeV/nucl and $\approx$ PeV. The lower part of the energy range corresponds to the so-called *anomalous cosmic rays* (ACR). These particles with low charge values can easily penetrate through the Earth's magnetic field and after charge-exchange with neutrals of the Earth's atmosphere increase their charge states to full stripping. As a result of satellite experiments it was shown, that these secondary particles can become trapped in the geomagnetic field and form a radiation belt. Below a short survey of the experimental results which led to this discovery is given.

Keywords: Anomalous Cosmic Rays

1. Introduction

The term *anomalous cosmic rays* was first introduced in 1973 after the discovery of a local maximum at $\approx$ 10 MeV/nucl in the energy spectrum of such cosmic ray elements as ^{4}He and ^{16}O [1,2]. This maximum appeared during solar activity minimum (in 1976-1977) at about 10 MeV/nucl, i.e. at energies between those of particles of solar origin - solar energetic particles (<10 MeV/nucl) and traditional galactic cosmic rays (>100 MeV/nucl).

After the discovery of the GCR anomalous component Fisk, Kozlovsky and Ramaty [3] suggested a theory describing their origin. According to their hypothesis ACR are neutral atoms of the Local Interstellar Medium (LISM) which penetrate inside the heliosphere where they are ionised by solar ultraviolet radiation or due to charge-exchange with ions of the solar wind; then they are picked up by the solar wind and carried away towards the heliopause, where they are accelerated to energies of $\approx$ 10 MeV/nucl, and finally return back to the Sun. Later it was shown, that this process can be multiple [4]. The charge state of ACR ions can be 1+ or >2+.

The suggested mechanism of ACR origin and acceleration [3] assumes a relative increase of intensity for the elements with a high ionisation potential, whereas for elements with a low ionisation potential (e.g. for Mg, Si, Fe) there should be no

---

*E-mail: panasyuk@sinp.msu.ru

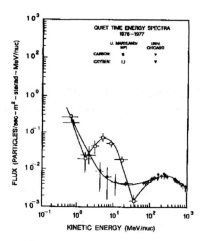

Fig. 1. The ACR energy spectra: He, C, N, O , according to the Voyager-2 observations at ≈23 AE in 1987[1]

anomalous increase of the flux.

Fig.1 shows the energy spectra of He, C, N, and O observed by Voyager-2 at the distance of 23 AU during the year of minimum solar activity. The maximum of ACR fluxes is revealed at energies <50 MeV/nucl. At larger energies GCR particles start to dominate.

It was experimental proof of ACR ions having charge states close to +1, that could serve as the final argument in favour of the above described model of ACR propagation and acceleration in the heliosphere.

The most convincing proof that ACR have charge states close to 1+ were the results of a series of experiments which used the effect of ion separation by the Earth's magnetic field. (see e.g. [5,6,7]). This technique was based on the comparison of ^{16}O fluxes, observed by the IMP-8 satellite in the interplanetary medium and simultaneous measurements at low altitudes (below 350 km) on satellites of the Cosmos series. Detailed analysis of these results is given in [5]. Unambiguous determining of the charge state of ^{16}O was the key result of these experiments and proof of the validity of the Fisk, Kozlovsky and Ramaty hypothesis. Fig.2[3] shows the results of comparison of the ^{16}O ion data (measured on Cosmos and IMP-8 satellites) and model calculations, based on the penetration ^{16}O ions with charge values Q=+1 and Q=+8 into the inner magnetosphere. The mean charge state for ^{16}O was found to be $Q=0.9^{+0.3}_{-0.9}$

Studies of ACR are very important for cosmic ray physics. It is known that while propagating through the interstellar medium cosmic rays (ions with energies above ≈ 10 MeV/nucl) soon become fully stripped. In order to lose their orbital electrons the ions need to travel through just several tens of mg/cm^2 of matter. Therefore, the presence of orbital electrons in the low energy cosmic ray component atoms can serve as evidence that their source is located in the direct vicinity of the solar

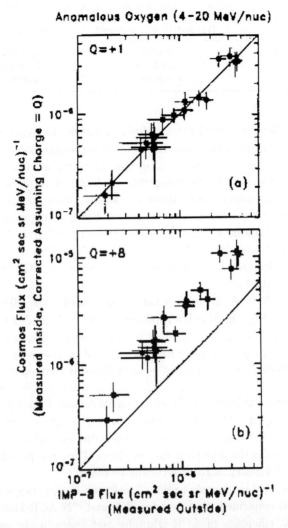

Fig. 2. Comparison of anomalous oxygen intensities measured inside the magnetosphere (Cosmos data) and outside the magnetosphere (IMP-8 data) for supposed ^{16}O charged states of +1 and +8[5]

system. Losses of O^+ ions associated with charge-exchange processes ($O^+ \to O^{n+}$), which impose limitations on the life times of these particles in the process of their propagation, were studied in [8]. The authors showed that, for density of H neutrals ≈ 0.1 cm^{-3} and upper limit of the mean charge state of O^+ ions $<Q> = 1.6$, their maximum propagation distance should not exceed 0.2 pc for O^+ at energies of ≈ 10 MeV/nucl. I.e., this actually means, that the sources of ACR are located somewhere in the near-by regions of the Universe - the LISM. Propagating inside the solar system ACR ions (which have significant momentum) due to their small charge state, can reach the vicinities of planets, which have magnetic fields.

The first measurements of O ions in the vicinity of the Earth were made in

Table 1. The relative abundances of ACR according to SAMPEX data

Ratio	Trapped 16-45 MeV/nucl	Interplanetary
C/O	≈0.0004	0.014(0.009)
N/O	0.09(0.01)	0.19(0.03)
Ne/O	0.04(0.01)	0.06(0.02)

1973-74 on the Skylab orbital station (orbit altitude 435 km) by Biswas et al. (see e.g. survey [9]) using solid state Lexan Poly-carbonate detectors. It was these measurements that permitted to point out that O ion fluxes recorded in the Skylab orbit (exceeding those in the interplanetary medium) could be associated with O ions having a small charge state. However, at that time no proof was obtained, that these are in fact particles trapped in the Earth's magnetic field. The same pioneering works of Biswas et al. presented data on the relative abundance of N; C; Ne. In particular, it was shown that O/C=4.7 ± 0.1, which served as a convincing indication of the close relations of these particles with the ACR fluxes observed in the interplanetary medium.

Important result of ACR studies carried out during the 90's was proof that ions can be trapped by the Earth's geomagnetic field. This idea was suggested by Blake and Friesen in 1977 [9]. According to this theory a singly-ionised ACR ion, penetrating inside the geomagnetic field, is stripped by the residual atmosphere at altitudes of ≈ 300 km, becomes multiply-charged and is trapped by the geomagnetic field.

This theory was confirmed in the experiments onboard satellites of the Cosmos series [10]. The trapped ^{16}O were actually recorded using angular distributions of particle tracks in solid state detectors; and their flux exceeded by a factor of hundreds the ^{16}O fluxes in the interplanetary medium. Thus, the possibility of studying ACR in the direct vicinity of the Earth was revealed.

Starting from 1992 large-scale tudies of penetrating and trapped ACR began on SAMPEX. It was confirmed, that besides ^{16}O, and ^{14}N ACR also contain Ne and Ar. The different efficiency of ACR trapping and losses in the geomagnetic field effect their energy spectra: they become softer in comparison to the interplanetary ones [11]. As a consequence, their relative composition also changes. Table 1 shows the relative abundances for the individual components of ACR for trapped and interplanetary ions according to SAMPEX data [5]. The discrepancies are obvious.

Measurements of ACR during the last two solar activity cycles show that this GCR component is extremely sensitive to solar modulation. The flux of anomalous ^{16}O with ≈ 10 MeV/nucl in the interplanetary medium varies within a factor of ≈ 100. Temporal variations of trapped ACR were studied for the first time on satellites of the Cosmos series (see Fig.3, [5]). It turned out, that the behaviour of trapped anomalous ^{16}O is similar. This means, that the life time of the trapped component is relatively small - does not exceed the characteristic times of interplanetary ACR flux variations.

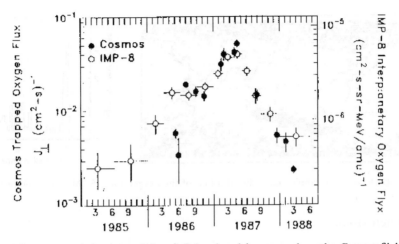

Fig. 3. Time history of the trapped flux (left-hand scale) measured on the Cosmos flights and the quite-time interplanetary 5-11 MeV/nucleon oxygen flux (right-hand scale) measured by the instrument on IMP-8 [10].

On the basis of Cosmos data a model of the spatial distribution of trapped ACR particles was built. It was revealed, that the ACR belt is localized slightly eastwards of the South-Atlantic Anomaly (see Fig.4.) with maximum intensities at L-shells of about 2.2-2.5. The results of this calculation model [12] were later confirmed by direct observations of trapped ACR particles on the SAMPEX satellite (see Fig.5)

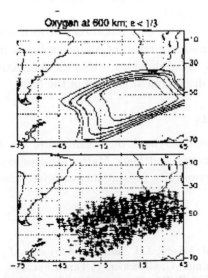

Fig. 4. Isoflux contours of the E >20 MeV/nucleon oxygen at 600 km (top panel) and MonteCarlo simulations of the distribution produced by sampling along the SAMPEX satellite orbit [12]

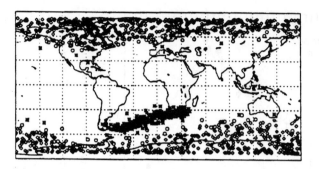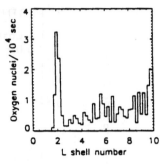

Fig. 5. Observed geographic and L-distributions of the oxygen ions (SAMPEX satellite data) [11]

2. Conclusions

Satellite studies of anomalous cosmic rays carried out during the 70's-90's led to the understanding that the most low-energy component of GCR observed in the heliosphere consists of particles which originate in the LISM - the part of the Galaxy which is located close to us, and the limited time of their propagation and acceleration inside the heliosphere is responsible for their low charge state. These particles can penetrate inside the magnetosphere, and, interacting with the atmosphere, are transformed into ions with larger charge states. In this way the conditions for their trapping inside the magnetic field are created. The radiation belt, formed by ACR particles, is, essentially different from those known previously for the following reasons:

(1) The source of these particles is not the Sun and not the Earth's ionosphere, but matter from the LISM;
(2) The main mechanism of the new radiation belt formation is trapping of multi-charged ions, produced as a result of charge-exchange of singly charged ions in the ionosphere, not radial diffusion across the magnetic field lines, typical for the radiation belts discovered at the end of the 50's.

3. Acknowledgements

The author sincerely wishes to thank the organizers of the 2nd International Workshop on Matter, Antimatter and Dark Matter, and especially Professor Roberto Battiston for the opportunity to present this tutorial lecture and to publish it in these Proceedings.

References

1. M.Garcia-Munoz, G.M. Mason, J.H. Simpson, *Astrophys. J.*, **182**, L81-L84 (1973)
2. D.O. Hovestadt, 0. Valmer, G. Gloeckler, C. Fan, *Phys. Rev. Lett* **31**, 650-667 (1973).
3. L.A. Fisk, B. Kozlovskiy, R. Ramaty *Astrophys. J.* **190**, L35-L38 (1974).
4. R.A. Mewaldt, R.S Selesnick,J.R. Cummings, E.C. Stone, T.T. Von Rosenvinge *Astrophys. J* **466**, L43-L46 (1996).

5. J.H. Adams,M. Garcia-Munoz,N.L. Grigorov,B. Klecker, M.A. Kondratyeva,G.M. Mason,DE. McGuire,R. Mewaldt, M.I. Panasyuk, Ch.A. Tretyakova, A.J. Tylka, D.A. Zhuravlev, *Astrophys J.* **375**, L45-L48 (1991)
6. M. Panasyuk, *Proc. of the 23th Int. Cosmic Ray Conference* (Invited, rapporteur and highlight papers), 455-463 (1993).
7. B. Klecker, *Space Science Reviews* **72**, 419-430 (1995).
8. J.H. Adams, Leising,*Proc. 22 Intern. Cosmic Ray Conf. (Dublin)* **3**, 304 (1996).
9. S. Biswas, *Space Science Reviews* **75**, 423-451 (1996).
10. Blake, J.B. and Friesen , L.M. *Proc. of the 15th ICRC* **2**, 341-346 (1977).
11. N.L. Grigorov, M.A. Kondratyeva, M.I. Panasyuk, Ch.A. Tretyakova, J.H. Adams, J.B. Blake, M. Shultz, R.A. Mewaldt, A.J. Tylka, *Geophys. Res. Letters* **18**, 1959-1962 (1991).
12. R.S. Selesnik, A.C. Cummings, J.R. Cummings, R.A. Mewaldt, E.C. Stone, T.T. Von Rosenvinge, *J.Geophys. Res* **100**, 9503-9509 (1995).
13. A. Tylka, Preprint Cosmic Ray Section E.O.Hulbert Center for Space Research, Naval Research Laboratory, 10 p.p. Proc. of the 23th Intern. Cosmic Ray Conf. 5 (1993)

BIOLOGICAL EFFECTS OF COSMIC RADIATION IN LOW-EARTH ORBIT

MARCO DURANTE*

Dipartimento di Scienze Fisiche, Università Federico II
Monte S. Angelo, Via Cintia, 80126 Napoli, Italy.

Space exploration poses health hazards to the crews of manned missions. Exposure to cosmic radiation and loss of bone density are considered the two most important risk factors for long-term missions. Stochastic risk deriving from cosmic radiation exposure can be estimated by physical dosimeters, using appropriate conversion factors. Recent measurements of space radiation fluence and energy spectra will improve current estimates. Biological dosimetry can be used as a tool to determine the risk directly from biological damage. Chromosomal aberrations in astronauts' peripheral blood lymphocytes have been used as a biomarker of cancer risk. In this paper we will also discuss countermeasures to radiation damage, focusing on the problem of shielding in space.

Keywords: Cosmic Radiation, biological effects

1. Introduction

Manned space exploration started about four decades ago, and space radiation was soon recognized as a health hazard for the crew. More than 300 astronauts have been exposed to cosmic radiation during manned space missions, and the International Space Station (ISS) program will add substantially to this number and significantly increase average lifetime doses. The methods employed to estimate the space radiation risk have not changed significantly over this long period of time. Basically, absorbed dose is measured using physical dosimeters, such as thermoluminescence detectors (TLDs), and converted in equivalent dose using measured linear energy transfer (LET) spectra. Risk is derived by equivalent doses based on databases of individuals exposed to radiation on Earth, such as the A-bomb survivors. Uncertainties associated to this method are still quite large, due to the poor knowledge of the biological effects of heavy ions, and to the specific exposure conditions in space (i.e., the effect of microgravity on radiation response).

NASA limits an astronauts' career radiation exposure to a projected risk of 3% excess cancer fatality [1]. In addition, short-term dose limits (30-days and 1-year) are imposed to prevent deterministic effects. A detailed analysis of astronauts' exposures in all missions performed so far shows that the excess cancer risk never

---

*E-mail: durante@na.infn.it

exceeds 0.5% [2]. However, these figures might change with the long-term missions in low Earth orbit (LEO) on the ISS, and especially with a manned Mars mission.

Two main problem are associated to space radiation protection: a) reducing uncertainties on risk estimates, using either ground-base radiobiological experiments or in-flight measurements of biological effects, and b) to develop appropriate countermeasures to the cosmic radiation exposure. This two issues will be briefly reviewed in this paper.

2. Space Radiation Dosimetry

Space radiation consists primarily of high-energy charged particles, such as protons, helium, and heavier ions (HZE particles), originating from several sources, including galactic cosmic radiation (GCR), solar flares and Van Allen belts. High-energy electromagnetic radiation and neutrons have been measured on spacecrafts, too. Space radiation is very penetrating, and shielding can reduce, but not eliminate, crew exposure to ionizing radiation. The van Allen belts are formed mostly of electron and protons trapped in closed orbits in the Earth's geomagnetic field. Both GCR and solar particle radiation consist mostly of protons, with a smaller components of helium and heavier nuclei. LEO crews are exposed mostly to the trapped particles in the geomagnetic field, and exposure depends strongly on orbit altitude and inclination. Outside magnetoshpere, a chronic low-dose exposure will be provided by cosmic rays, and possible acute high-dose exposure can be induced by a solar particle event (SPE). Particle fluxes are generally modulated by solar activity, which has a fairly regular 11-years cycle.

During mission planning, it is important to have reliable estimates of the absorbed dose for crewmembers, of the projected equivalent doses, and of cancer risks. The calculation of mission exposure and cancer risks starts from the information about mission trajectory and spacecraft shielding. Transport codes for GCR and trapped radiation are used to generate the mission radiation spectrum. Human anatomical models are used to calculate body-self shielding, and fluxes in single organs. From the absorbed doses, dose equivalent and cancer risk are estimated by appropriate coefficients.

Dosimetry of space radiation has been performed in the last 30 years, and results are available for a number of different space situations. In most cases, TLDs have been used for determination of absorbed dose of sparsely ionizing radiation, while HZE particles have been detected by solid state nuclear track detectors (SSNTDs). TLDs (LiF or Al_2O_3) are used as personal dosimeters, and astronauts are supposed to wear the badges all the time. Dosimeter packets with TLDs and SSNTDs (CR-39, cellulose nitrate and Lexan polycarbonate) are also positioned in different points of the spacecraft. Many reviews of such data are available [1,3-5]. Active radiation monitors such as tissue equivalent proportional chambers (TEPCs) have been flown on a number of missions [6-7] to measure absorbed dose and LET spectra. Analysis of high temperature emissions from TLDs has also been used to provide simultaneous measurements of dose and average LET [8-9]. Neutron spectrometers have also been flown on Shuttle and Mir.

The comparison between dose measurements and models are generally in reasonable good agreement, although some improvements are necessary [10]. Models of GCR are sufficiently accurate to provide accurate estimates of equivalent doses, although some discrepancies were found between LET spectra calculated by HZETRN and measurements from TEPC on the Space Shuttle [11]. The AP-8 model for trapped protons substantially overestimated doses on Shuttle and MIR [12] and the agreement between measured and predicted trapped proton LET spectra was quite poor. AP-8 predicts a higher flux of low LET particles than measured by TEPC, and consequently higher dose rates and smaller quality factor.

Table 1. Radiation dose equivalents for different missions. Data are from radiation measurements, after correction for quality factors, except for ISS and Mars, where we present the best estimates. Maximum values are reported in parenthesis. Background Equivalent Radiation Time (BERT) is based on the UK average of 2.2 mSv/year. Data courtesy of Dr. Francis Cucinotta (NASA Johnson Space Center).

Program	aver. alt. (km)	Inclin. (degrees)	Crew	Dose-rate (mSv/day)	Tot. dose (cSv)	BERT (years)
Gemini	454 (1370)	30	20	0.87 (4.7)	0.053 (0.47)	0.24 (2.0)
Apollo	-	-	33	1.3 (3.9)	1.22 (3.3)	5.5 (15)
Skylab	381 (435)	50	9	1.2 (2.1)	7.2 (17.0)	32 (77)
STS (<450 km)	337	28.5	207	0.23 (0.4)	0.21 (0.71)	1 (3.2)
STS (>450 km)	570	28.5	85	3.2 (7.7)	2.65 (7.8)	12 (35)
STS/MIR	341 (355)	51.6	4	0.72 (1.0)	9.9 (14.0)	45 (63)
ISS	360 (450)	51.6	280	0.5 (0.1)	8 (18)	36 (82)
MARS	-	-	4 (8)	1.5 (2.0)	40 (120)	182 (545)

Improvement of current models of space radiation require a better knowledge of the radiation environment in the free space, and its interaction with shielding materials. A considerable improvement in radiation measurements in space is expected by the Alpha Magnetic Spectrometer (AMS-II), slated for launch in September 2003. This advanced version of AMS-I, that flew on STS-91, will measure the energy spectrum of particles from hydrogen to iron in the energy range from 0.07 to 1.4 GeV/n [13]. ESA will also provide two detector telescopes (DOSTEL) to measure time-resolved particle count and LET spectra [14]. A number of active radiation monitors will provide on ISS data for use in operational monitoring, such as a warning in case of SPE [13].

Based on absorbed dose measurements aboard spacecrafts and transport models, a number of mission scenarios can be simulated and relative risk estimated [15]. Some examples are given in Table 1, and compared with Background Equivalent

Radiation Time (BERT), i.e. with the time on Earth providing the same equivalent dose [16]. In BERT calculations, we assumed the natural background based on UK average of 2.2 mSv/year.

3. Biological Dosimetry

The weak point in the risk estimates shown in Table 1 rely on the quality factors associated to radiation of different quality, and to the risk coefficients. All these coefficients bear large uncertainties [17]. It has been estimated that the additional cancer risk induced by exposure to space radiation could be from 4 to 15 times[18] larger or smaller than calculated according to best currently available data and methods. The desired uncertainty goal to be reached within year 2023 is set to ±50% in the NASA Strategic Plan [19].

Reducing the uncertainty on risk estimates requires ground-base radiobiological experiments, as well as biological indicators of space radiation risk, which could directly provide risk estimates. A common biomarker of radiation risk are chromosomal aberrations (CA), usually measured in the peripheral blood lymphocytes (PBL) of exposed individuals [20]. Evidence that CA in PBL are positively correlated with cancer risk has been recently obtained in a large cohort study, performed by the European Study Group on Cytogenetic Biomarkers and Health (ESCH), where a group of 3541 healthy subjects from 5 European countries were screened for CA over a period of three decades [21].

Biodosimetry based on CA in PBL requires blood samples before and after the space missions. Chromosomes can be visualized at the first mitosis following *in vitro* growth stimulation of PBL, or by premature chromosome condensation (PCC). The yield of CA in PBL can be measured by different techniques, including Giemsa-staining or fluorescence *in situ* hybridization (FISH). Measurements performed so far prove that a significant increase in the frequency of CA is observed after long-term space missions [22-27].

Data on CA induction can be used for different purposes. First, using appropriate dose-response curves measured *in vitro* with reference radiation, the equivalent dose absorbed during the mission can be estimated and compared with the equivalent dose measured by physical dosimeters. Examples of such comparisons are provided in Table 2 [27]. Despite the large uncertainties in both biological and physical measurements, data agree quite well, suggesting that the method for risk estimates based on currently available data is fairly accurate. Second, the relative increase in measured CA can be used to estimate directly the cancer risk, using the ESCH database [28]. Estimates are provided in Table 3 for crew of the Russian Space Station Mir. An overall relative risk around 1.2 is estimated for crewmembers after long-term Mir missions.

Table 2. Comparison of physical and biology dosimetry in six astronauts involved in Mir or ISS 3-months space missions. Dose measured by TLD is corrected to provide the dose at the blood forming organs (BFO), and then multiplied by the average quality factor Q to provide the equivalent dose. Radiation quality factor Q is estimated by TEPC measurements of LET spectrum, using the Q-LET relationship recommended by ICRP and NCRP[1]. Biological dose is estimated by measurements of chromosomal aberrations in post-flight blood samples using *in vitro* calibration curves obtained on Earth exposing the pre-flight astronauts' blood to reference γ-rays. Data from ref. 27.

Crew numb.	Dose (TDL) (mGy)	Dose (BFO) (mGy)	Q (TEPC)	Eq. dose at BFO (mSv)	Biol. dose by CA (mSv)
1	31	28	2.5	71	44±36
2	38	34	2.3	81	100±32
3	57	51	2.4	125	102±23
4	42	38	2.6	100	147±46
5	34	32	2.6	82	31±73
6	44	41	2.6	106	137±41
Total		228		567	561
Average Q			2.5	2.48	2.46

Table 3. Estimates of risk from cytogenetic analyses in different space missions. The fraction of aberrant cells, which include all types of aberrations, are evaluated as average values on the number of crewmembers specified in column 2. Data from references 24 and 27 were obtained using FISH with a combination of whole-chromosome probes specific for chromosomes 1, 2, 4, or 5, and then scaled to the whole genome. Data from references 22, 23, and 25 were obtained from Giemsa-stained specimens. For post-flight data (column 5), sampling time after returning on Earth was between 0 and 180 days. Ratios of fraction of aberrant cells reported in column 5 and 4 are given in column 6. Relative risk is reported in column 7, and is based on the ESCH database[28]. Physical doses reported in column 8 were measured by TLD. In this column, dose ranges are reported when measured doses differ for the crewmembers studied.

Space mission (ref.)	Crew	Fligth duration (days)	% ab. cells (before flight)	% ab. cells (after flight)	Freq. ratio	Relat. risk	Dose (mGy)
MIR 18 (24)	2	115	4.4	8.9	2.0	1.23	42
Antares (22)	3	180	1.1	2.0	1.9	1.17	90
EuroMir (23)	6	120-198	0.6	2.0	3.5	1.57	61-101
Mir (25)	22	100-250	1.7	2.4	1.4	1.09	30-58
ISS/Mir (27)	6	115-144	1.4	2.3	1.6	1.14	36-67
Total	39	100-250	1.8	3.5	2.1	1.2	30-101

4. Countermeasures

For terrestrial radiation workers, protection against radiation exposure can be provided through shielding of the radiation source. In extra-terrestrial space, shielding is effective against trapped protons, but its efficiency is poor against GCR penetration. Indeed, high-energy particle radiation in space is very penetrating and GCR produce a large number of secondary particles, including neutrons, generated by nuclear interactions with the nuclei in the shield. These particles have generally lower energy, but can have higher quality factors than incident primary cosmic particles.

A number of computer calculations performed by NASA showed that the shield performance is dependent upon shield material and thickness, as well as incident beam energy and charge [29-30]. Results of calculations suggest that the biological effectiveness of GCR can be increased behind thin shields made by high atomic number materials, because of nuclear fragmentation of heavy-ion projectiles. Recently, direct biological measurements of shield performance have been performed at the HIMAC accelerator in Japan and at the AGS accelerator at the Brookhaven National Laboratories in USA [31-32]. These measurements focus on 200-1000 MeV/n ^{56}Fe beams shielded with PMMA (lucite) or aluminum. Biological endpoints are chromosomal aberrations, DNA double-strand breakage, and cell killing.

Fig. 1 and Fig. 2 show results for the induction of CA in PBL exposed *in vitro* at the HIMAC accelerator to iron ions shielded with either PMMA or aluminum. When frequency of CA is plotted vs. the dose at the sample position (Fig. 1), no significant differences are detected between shielded and unshielded 500 MeV/n Fe-beams. However, plotting the data as a function of the particle fluence incident on the shield (Fig. 2), we observe a significant increase in the biological damage per incident particle produced by the shielding. These data provide a first direct evidence that shielding space radiation can, under certain circumstances, increase rather than reduce the radiation risk for the crew.

5. Conclusions

Space radiation is a serious, but potentially solvable problem for human space colonization. Although the uncertainties about radiation risk are still quite large, biological dosimetry suggest that today estimates are sound and can be used as guidelines for protection of crews. Further improvements in risk estimates are expected from AMS detectors, and ground-based biological experiments. Radiation risk can be reduced by shielding, but nuclear fragmentation complicates the problem in space. More data are now available which will allow optimized design of shielding in spacecrafts.

Acknowledgements

Research activity in space biological dosimetry performed by the author is supported by EU- INTAS grant 99-00214, and University Space Research Association

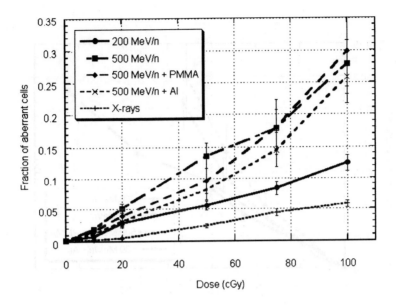

Fig. 1. Fraction of human lymphocytes displaying aberrations in prematurely condensed chromosomes 1, 2 or 4 after exposure to accelerated iron ions plotted as a function of dose. The 500 MeV/n iron beam was shielded with either 56 mm PMMA (lucite) or 30 mm Al. This shielding reduces the residual range to the same value of the unshielded 200 MeV/n iron beam.

(Houston, TX). Research about shielding performance under heavy-ion bombardment is generously supported by the Italian Space Agency (ASI), grant I/R/034/01. The author is grateful to Dr. Francis Cucinotta (NASA Johnson Space Center) for useful comments and for providing the data displayed in Table 1. Most of the information concerning physical dosimetry aboard spacecrafts come from the late Dr. Gautam Badhwar (NASA Johnson Space Center). His premature departure is a tragic loss for his family, friends, and for the forthcoming space missions. This review is dedicated to his memory. I am also grateful to Prof. Giancarlo Gialanella (University Federico II, Naples, Italy) for his continuos encouragement and support of space radiation activities in Italy. Finally, I thank Dr. Antonio Ereditato for introducing me to the mysteries of LateX.

References

1. NCRP, *Radiation Protection Guidance for Activities in Low – Earth Orbit* (NCRP Report n. 132, National Council on Radiation Protection and Measurements, Bethesda, MD, 2001).
2. F.A. Cucinotta, *Physica Medica* **17** (suppl. 1), 5 (2001).

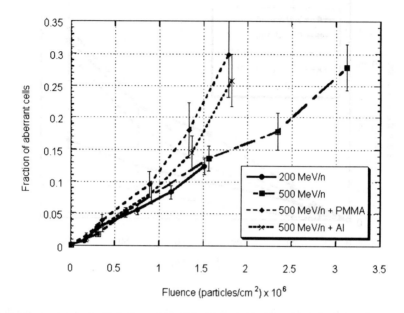

Fig. 2. Fraction of human lymphocytes displaying aberrations in prematurely condensed chromosomes 1, 2 or 4 after exposure to accelerated iron ions plotted as a function of fluence of particles incident on the shield. Shielding as in Fig.1.

3. R. Beaujean, J. Kopp and G. Reitz, *Radiat. Prot. Dosim.* **85**, 223 (1999).
4. G. D. Badhwar, *Radiat. Res.* **157**, 69 (2002).
5. G. Reitz, R. Beujean, J. Kopp, M. Luszik-Bhadra, W. Heinrich and K. Strauch, in *Exploring Future Research Strategies in Space Radiation Sciences*, ed. H.J. Majima and K. Fujitaka (Iryokagakusha, Tokyo, 2000) p. 71.
6. J.L. Shinn, G.D. Badhwar, M.A. Xaspos, F.A. Cucinotta and J.W. Wilson, *Radiat. Meas.* **30**, 19 (1999).
7. G.D. Badhwar, V.V. Kushin, Y.A. Akatov and V.A. Myltseva, *Radiat. Meas.* **30**, 415 (1999).
8. H. Yasuda, *Health Phys.* **80**, 576 (2001).
9. M. Noll, N. Vana, W. Schoner and M. Fugger, *Radiat. Prot. Dosim.* **85**, 283 (1999).
10. P. Denkins, G.D. Badhwar, V. Obot, B. Wilson and O. Jejelwo, *Acta Astronaut.* **49**, 313 (2001).
11. F.A. Cucinotta, J.W. Wilson, J.L. Shinn and R.K. Tripathi, *Adv. Space Res.* **21**, 1753 (1998).
12. G.D. Badhwar, *Radiat. Meas.* **30**, 401 (1999).
13. G.D. Badhwar, *Physica Medica* **17** (suppl. 1), 287 (2001).
14. G. Reitz, *Physica Medica* **17** (suppl. 1), 283 (2001).
15. NCRP, *Guidance on Radiation Received in Space Activities* (NCRP Report n. 98,

National Council on Radiation Protection and Measurements, Bethesda, MD, 1989).
16. NCRP, *Research Needs for Radiation Protection* (NCRP Report n. 117, National Council on Radiation Protection and Measurements, Bethesda, MD, 1993).
17. F.A. Cucinotta, W. Schimmerling, J.W. Wilson, L.E. Peterson, G.D. Badhwar, P.B. Saganti and J.F. Dicello, *Radiat. Res.* **156**, 682 (2001).
18. National Academy of Sciences, *A Strategy for Research in Space Biology and Medicine in the New Century* (National Academy Press, Washington DC, 1998).
19. NASA Life Sciences Division, *Strategic Program Plan for Space Radiation Health Research* (NASA, Washington DC, 1998).
20. M. Durante, *La Rivista del Nuovo Cimento* **19**, 12 (1996).
21. L. Hagmar, S. Bonassi, U. Strömberg, A. Brogger, L. Knudsen, H. Norppa, and C. Reuterwall, *Cancer Res.* **58**, 4117 (1998).
22. I. Testard, M. Ricoul, F. Hoffschir, A. Fluty-Herard, B. Dutrillaux, B.S. Fedorenko, V. Gerassimenko, and L. Sabatier, *Int. J. Radiat. Biol.* **70**, 403 (1996).
23. G. Obe, I. Johannes, C. Johannes, K. Hallman, G. Reitz, and R. Facius, *Int. J. Radiat. Biol.* **72**, 727 (1997).
24. T.C. Yang, K. George, A.S. Johnson, M. Durante and B.S.Fedorenko, *Radiat. Res.* **148** (suppl. 1), 17 (1997).
25. B. Fedorenko, S. Druzhinin, L. Yudaeva, V. Petrov, Y. Akatov, G. Snigiryova, N. Novitskaya, V. Shevchenko and A. Rubanovich, *Adv. Space Res.* **27**, 355 (2001).
26. I. Testard and L. Sabatier, *Mutat. Res.* **430**, 315 (1999).
27. K. George, M. Durante, H.-L. Wu, V. Willingham, G.D. Badhwar and F.A. Cucinotta, *Radiat. Res.* **156**, 731 (2001).
28. M. Durante, S. Bonassi, K. George and F.A. Cucinotta, *Radiat. Res.* **156**, 662 (2001).
29. J.W. Wilson, F. A. Cucinotta, M.H.Y. Kim and W. Schimmerling, *Physica Medica* **17** (suppl. 1), 67 (2001).
30. J.W. Wilson, F.A. Cucinotta, S. A. Thibeault, M.H.Y. Kim, J.L. Shinn and F.F. Badavi, in *Shielding Strategies for Human Space Exploration*, ed. J.W. Wilson et al. (NASA CP-3360, Hampton, Virginia, 1997) p. 109.
31. M. Durante, *Physica Medica* **17** (suppl. 1), 269 (2001).
32. M. Durante, in *Exploring Future Research Strategies in Space Radiation Sciences*, ed. H.J. Majima and K. Fujitaka (Iryokagakusha, Tokyo, 2000) p. 79.

NEUTRINOS AS DARK MATTER CANDIDATES

ANTONIO MASIERO*

Dipartimento di Fisica, Università di Padova
via F. Marzolo, 8 I-35131 Padova, Italy

SILVIA PASCOLI[†]

Sissa and INFN Trieste
via Beirut 2-4
I-34014 Trieste, Italy

We review some properties of neutrinos in relation to the dark matter and baryogenesis problems.

Keywords: Neutrinos; Dark matter; Particle Physics.

1. Introduction

The electroweak standard model (SM) is now approximately thirty years old and it enjoys a full maturity with an extraordinary success in reproducing the many electroweak tests which have been going on since its birth. Not only have its characteristic gauge bosons, W and Z, been discovered and also has the top quark been found in the mass range expected by the electroweak radiative corrections, but the SM has been able to account for an impressively long and very accurate series of measurements. Indeed, in particular at LEP, some of the electroweak observables have been tested with precisions reaching the per mille level without finding any discrepancy with the SM predictions. At the same time, the SM has successfully passed another very challenging class of exams, namely it has so far accounted for all the very suppressed or forbidden processes where flavor changing neutral currents (FCNC) are present.

By now we can firmly state that no matter which physics should lie beyond the SM, necessarily such new physics has to reproduce the SM with great accuracy at energies of $O(100$ GeV$)$.

And, yet, in spite of all this glamorous success of the SM in reproducing the impressive set of experimental electroweak results, we are deeply convinced of the existence of new physics beyond this model. We see two main classes of motivations pushing us beyond the SM.

---

*E-mail: antonio.masiero@pd.infn.it
[†]E-mail: pascoli@sissa.it

First, we have theoretical "particle physics" reasons to believe that the SM is not the whole story. The SM does not truly unify the elementary interactions (if nothing else, gravity is left out of the game), it leaves the problem of fermion masses and mixings completely unsolved and it exhibits the gauge hierarchy problem in the scalar sector (namely, the scalar higgs mass is not protected by any symmetry and, hence, it would tend to acquire large values of the order of the energy scale at which new physics sets in). This first class of motivations for new physics is well known to particle physicists. Less familiar is a second class of reasons which finds its origin in some relevant issues of astroparticle physics. We refer to the problems of the solar and atmospheric neutrino deficits, baryogenesis, inflation and dark matter (DM). In a sense these aspects (or at least some of them, in particular the solar and atmospheric neutrino problems and DM) may be considered as the only "observational" evidence that we have at the moment for Physics beyond the SM.

2. Neutrinos and the Dark Matter Problem

Let's define Ω (for a review see [1] and [2]) as the ratio between the density ρ and the critical density $\rho_{crit} = \frac{3H_0^2}{8\pi G} = 1.88\, h_0^2 \times 10^{-29} \text{g cm}^{-3}$ where H_0 is the Hubble constant, G the gravitational constant:

$$\Omega = \frac{\rho}{\rho_{crit}}. \tag{1}$$

The Ω_{lum} due to the contribution of the luminous matter (stars, emitting clouds of gases) is given by:

$$\Omega_{\text{lum}} \leq 0.01. \tag{2}$$

First evidences of dark matter come from observations of galactic rotation curves (circular orbital velocity vs. radial distance from the galactic center) using stars and clouds of neutral hydrogen. These curves show an increasing profile for little values of the radial distance r while for bigger ones it becomes flat, finally decreasing again. According to Newtonian mechanics this behavior can be explained if the enclosed mass rises linearly with galactocentric distance. However, the light falls off more rapidly and therefore we are forced to assume that the main part of matter in galaxies is made of non-shining matter or dark matter (DM) which extends for a much bigger region than the luminous one. The limit on Ω_{galactic} which can be inferred from the study of these curves is:

$$\Omega_{\text{galactic}} \geq 0.1. \tag{3}$$

The simplest idea is to suppose that the DM is due to baryonic objects which do not shine. However Big-Bang nucleosynthesis (BBN) and in particular a precise determination of the primeval abundance of deuterium provide strong limits on the value of the baryon density [3] $\Omega_B = \rho_B/\rho_{crit}$:

$$\Omega_B = (0.019 \pm 0.001)h_0^{-2} \simeq 0.045 \pm 0.005. \tag{4}$$

1/3 of the BBN baryon density is given by stars, cold gas and warm gas present in galaxies. The other 2/3 are probably in hot intergalactic gas, warm gas in galaxies and dark stars such as low-mass objects which do not shine (brown dwarfs and planets) or the result of stellar evolution (neutron stars, black holes, white dwarfs). These last ones are called MACHOS (MAssive Compact Halo Objects) and can be detected in our galaxy through micro-lensing.

Anyway from cluster observations the ratio of baryons to total mass is $f = (0.075 \pm 0.002) h_0^{-3/2}$; assuming that clusters provide a good sample of the Universe, from f and Ω_B in (4) we can infer that:

$$\Omega_M \sim 0.35 \pm 0.07. \qquad (5)$$

Such value for Ω_M gets further supporting evidences from the evolution of the abundance of clusters and measurements of the power spectrum of large-scale structures.

Hence the major part of dark matter is non-baryonic [4]. The crucial point is that the SM does not posses any candidate for such non-baryonic relics of the Early Universe. Hence the demand for non baryonic DM implies the existence of new Physics beyond the SM. Non-baryonic DM divides into two classes [5,6]: cold DM (CDM), made of neutral heavy particles called WIMPS (Weakly Interacting Massive Particles) or very light ones as axions, hot DM (HDM) made of relativistic particles as neutrinos or even warm dark matter (WDM) with intermediate characteristics such as gravitinos.

In this talk we'll deal with the role of neutrinos as DM candidates. For a more general discussion on DM see the talk of P. Ullio at this meeting.

3. Neutrino masses and Mixing

The recent atmospheric neutrino data from SuperKamiokande and the solar neutrino data from SuperKamiokande and from SNO provide strong evidence of neutrino oscillations which can take place only if neutrinos are massive and mix. The parameters relevant in ν-oscillations are the mixing angles and the mass-squared differences which can be measured in atmospheric neutrinos, solar neutrinos, short-baseline and long-base line experiments:

i) in atmospheric neutrino experiments, to account for the deficit of the ν_μ flux expected towards the ν_e one from cosmic rays and its zenith dependence, it's necessary to call for $\nu_\mu \to \nu_\tau$ oscillations with:

$$\sin^2 2\theta_{atm} \geq 0.82 \qquad (6)$$
$$\Delta m^2_{atm} \simeq (1.5 - 8.0) \times 10^{-3} \text{ eV}^2 \qquad (7)$$

from SuperKamiokande data at 99% C.L. [7,8];

ii) the solar ν anomaly arises from the fact that the ν_e flux coming from the Sun is sensibly less then the one predicted by the Solar Standard Model: also this problem can be explained in terms of neutrino oscillations. The recent analysis of the SuperKamiokande data [9] and of the SNO data [10] favor the LMA (Large Mixing Angle) solution with:

$$\tan^2 \theta_\odot \simeq 0.15 \div 1.5 \quad (8)$$
$$\Delta m_\odot^2 \simeq (1.5 - 70) \times 10^{-5} \text{ eV}^2 \quad (9)$$

at 99% C.L.[11], even if the Small Mixing Angle ($\tan^2 \theta_\odot \sim 10^{-4}$) and the LOW ($\tan^2 \theta_\odot \sim 0.4 \div 4$) solutions cannot be excluded and the oscillations into sterile neutrinos are strongly disfavored;

iii) reactor [12], short baseline and long baseline experiments constrain further the parameters;

iv) finally the LSND experiment has evidence of $\bar{\nu}_\mu \to \bar{\nu}_e$ oscillations with $\Delta m^2_{\text{LSND}} \simeq (0.1 \div 2)$ eV^2, the KARMEN experiment has no positive results for the same oscillation and then restricts the LSND allowed region [13].

In the next future several long-base line experiments will be held to test directly ν-oscillations and measure the relevant parameters: K2K in Japan is already looking for missing ν_μ in $\nu_\mu \to \nu_\tau$ oscillations, MINOS (in US) and OPERA (with neutrino beam from CERN to Gran Sasso) are long-base line experiments devoted to this aim, which are under construction.

The tritium beta decay experiments are searching directly to detect the effective electron neutrino mass m_β and the present Troitzk [14] and Mainz [15] limits read $m_\beta \leq (2.5 \div 2.9)$ eV, there are perspectives to increase the sensitivity down to 0.4 eV (the KATRIN project[16]).

The $\beta\beta_{0\nu}$ decay predicted if neutrinos are Majorana particles depends on the value of the effective Majorana mass $|<m>|$:

$$|<m>| \equiv \left| \sum_i U_{ei}^2 m_i \right| \quad (10)$$

where m_i are neutrino masses and U_{ei} are the elements of the lepton mixing matrix. The present Heidelberg-Moscow[17] bound is $|<m>| \leq (0.35 \div 1)$ eV but in the next future there are perspectives to reach $|<m>| \sim 0.01$ eV (NEMO3, GENIUS, EXO, Majorana). If the searches for $\beta\beta_{0\nu}$ decay will give a positive result the nature of neutrinos will be established: they would be Majorana particles indistinguishable from their antiparticles. Furthermore the value of the effective Majorana mass $|<m>|$ would provide information on the neutrino mass spectrum and on the absolute value of neutrino masses. Combined with the information on m_1 from the tritium beta decay experiments, it would be also possible to get information on the Majorana CP-violating phases (present only if neutrinos are Majorana particles)[18].

From all these experiments we can conclude that neutrinos have masses and that their values must be much lower than the other mass scales in the SM.

4. Neutrino masses in the SM and beyond

The SM cannot account for neutrino masses: we cannot construct either a Dirac mass term as there's only a left-handed neutrino and no right-handed component, or a Majorana mass term because such mass would violate the lepton number and the gauge symmetry.

To overcome this problem, many possibilities have been suggested:

- within the SM spectrum we can form an $SU(2)_L$ singlet with ν_L using a triplet formed by two Higgs field H as $\nu_L \nu_L H H$. When the Higgs field H develops a vev this term gives raise to a Majorana mass term. However, this term is not renormalizable, breaks the leptonic symmetry and do not give an explanation of the smallness of neutrino masses;
- we can introduce a new Higgs triplet Δ and produce a Majorana mass term as in the previous case when Δ acquires a vacuum expectation value;
- however the most economical way to extend the SM is to introduce a right-handed component N_R, singlet under the gauge group, which couples with the left-handed neutrinos. The lepton number L can be either conserved or violated. In the former option neutrinos acquire a "regular" Dirac mass like for all the other charged fermions of the SM. The left- and right-handed components of the neutrino combine together to give rise to a massive four-component Dirac fermion. The problem is that the extreme lightness of neutrinos requires an exceedingly small neutrino Yukawa coupling of $O(10^{-11})$ or so. Although quite economical, we do not consider this option particularly satisfactory.

The other possibility is to link the presence of neutrino masses to the violation of L. In this case one introduces a new mass scale, in addition to the electroweak Fermi scale, in the problem. Indeed, lepton number can be violated at a very high or a very low mass scale. The former choice represents, in our view, the most satisfactory way to have massive neutrinos with a very small mass. The idea (see-saw mechanism, [19,20]) is to introduce a right-handed neutrino in the fermion mass spectrum with a Majorana mass M much larger than M_W. Indeed, being the right-handed neutrino a singlet under the electroweak symmetry group, its mass is not chirally protected. The simultaneous presence of a very large chirally unprotected Majorana mass for the right-handed component together with a "regular" Dirac mass term (which can be at most of $O(100 \text{ GeV})$) gives rise to two Majorana eigenstates with masses very far apart.

The Lagrangian for neutrino masses is given by:

$$\mathcal{L}_{\text{mass}} = -\frac{1}{2}(\bar{\nu}_L \ \overline{N}_L^c)\begin{pmatrix} 0 & m_D \\ m_D & M \end{pmatrix}\begin{pmatrix} \nu_R^c \\ N_R \end{pmatrix} + h.c. \qquad (11)$$

where ν_R^c is the CP-conjugated of ν_L and N_L^c of N_R. It holds that $m_D \ll M$. Diagonalizing the mass matrix we find two Majorana eigenstates n_1 and

n_2 with masses very far apart:

$$m_1 \simeq \frac{m_D^2}{M}, \quad m_2 \simeq M.$$

The light eigenstate n_1 is mainly in the ν_L direction and is the neutrino that we "observe" experimentally while the heavy one n_2 is in the N_R one. The key-point is that the smallness of its mass (in comparison with all the other fermion masses in the SM) finds a "natural" explanation in the appearance of a new, large mass scale where L is violated explicitly (by two units) in the right-handed neutrino mass term.

5. Thermal history of neutrinos

Let's consider a stable massive neutrino (of mass less than 1 MeV) (see for example [1]). If its mass is less than 10^{-4} eV it is still relativistic today and its contribution to Ω_M is negligible. In the opposite case it is non relativistic and its contribution to the energy density of the Universe is simply given by its number density times its mass. The number density is determined by the temperature at which the neutrino decouples and, hence, by the strength of the weak interactions. Neutrinos decouple when their mean free path exceeds the horizon size or equivalently $\Gamma < H$. Using natural units ($c = \hbar = 1$), we have that:

$$\Gamma \sim \sigma_\nu n_{e^\pm} \sim G_F^2 T^5 \tag{12}$$

$$\text{and } H \sim \frac{T^2}{M_{Pl}} \tag{13}$$

$$\text{so that } T_{\nu d} \sim M_{Pl}^{-1/3} G_F^{-2/3} \sim 1 \text{ MeV}, \tag{14}$$

where G_F is the Fermi constant, T denotes the temperature, M_{PL} is the Planck mass. Since this decoupling temperature $T_{\nu d}$ is higher than the electron mass, then the relic neutrinos are slightly colder than the relic photons which are "heated" by the energy released in the electron-positron annihilation. The neutrino number density turns out to be linked to the number density of relic photons n_γ by the relation:

$$n_\nu = \frac{3}{22} g_\nu n_\gamma, \tag{15}$$

where $g_\nu = 2$ or 4 according to the Majorana or the Dirac nature of the neutrino, respectively.

Then one readily obtains the ν contribution to Ω_M:

$$\Omega_\nu = 0.01 \times m_\nu(\text{eV}) h^{-2} \frac{g_\nu}{2} (\frac{T_0}{2.7})^3. \tag{16}$$

Imposing $\Omega_\nu h_0^2$ to be less than one (which comes from the lower bound on the lifetime of the Universe), one obtains the famous upper bound of $200(g_\nu)^{-1}$ eV on the sum of the masses of the light and stable neutrinos:

$$\sum_i m_{\nu_i} \leq 200(g_\nu)^{-1} \text{ eV}. \tag{17}$$

Clearly from Eq.(16) one easily sees that it is enough to have one neutrino with a mass in the (1 − 20) eV range to obtain Ω_ν in the 0.1 − 1 range of interest for the DM problem.

6. HDM and structure formation

Hence massive neutrinos with mass in the eV range are very natural candidates to contribute an Ω_M larger than 0.1. The actual problem for neutrinos as viable DM candidates concerns their role in the process of large scale structure formation. The crucial feature of HDM is the erasure of small fluctuations by free streaming: neutrinos stream relativistically for quite a long time till their temperature drops to $T \sim m_\nu$. Therefore a neutrino fluctuation in order to be preserved must be larger than the distance d_ν traveled by neutrinos during such interval. The mass contained in that space volume is of the order of the supercluster masses:

$$M_{J,\nu} \sim d_\nu^3 m_\nu n_\nu (T = m_\nu) \sim 10^{15} M_\odot, \tag{18}$$

where n_ν is the number density of the relic neutrinos, $M_\odot$ is the solar mass. Therefore the first structures to form are superclusters and smaller structures as galaxies arise from fragmentation in a typical top-down scenario. Unfortunately in these schemes one obtains too many structures at superlarge scales. The possibility of improving the situation by adding seeds for small scale structure formation using topological defects (cosmic strings) are essentially ruled out at present [21,22]. Hence schemes of pure HDM are strongly disfavoured by the demand of a viable mechanism for large-structure formation. To overcome this problem, CDM models (see the talk by P. Ullio in these proceedings), in which the main contribution to Ω is given by Cold Dark Matter, and ΛCDM or QCDM models have been proposed.

7. Dark energy, ΛCDM and xCDM or QCDM

The expansion of the Universe is described by two parameters, the Hubble constant H_0 and the deceleration parameter q_0:

i) $H_0 \equiv \dot{R}(t_0)/R(t_0)$, where $R(t_0)$ is the scale factor, t_0 the age of the universe at present epoch, and we have:

$$H_0 = 65 \pm 5 \text{ Km sec}^{-1} \text{ Mpc}^{-1} \quad (h = 0.65 \pm 0.05); \tag{19}$$

ii) $q_0 \equiv -\frac{\ddot{R}(t_0)}{H_0^2} R(t_0)$ says if the universe is accelerating or decelerating. q_0 is related to Ω_0 as:

$$q_0 = \frac{\Omega_0}{2} + \frac{3}{2} \sum_i \Omega_i w_i \tag{20}$$

where $\Omega_0 \equiv \sum_i \rho_i/\rho_{crit}$, Ω_i is the fraction of critical density due to the component i, $p_i = w_i \rho_i$ is the pressure of the component i, $\rho_{crit} = \frac{3H_0^2}{8\pi G} = 1.88 h^2 \times 10^{-29}$ g cm^{-3}.

Measurements of q_0 from high-Z Type Ia SuperNovae (SNeIa) [23,24] give strong indications in favor of an accelerating universe. Cosmic microwave background (CMB) data [25] and cluster mass distribution [26] seem to favor models in which the energy density contributed by the negative pressure component should roughly be twice as much as the energy of the matter, thus leading to a flat universe ($\Omega_{\text{tot}} = 1$) with $\Omega_M \sim 0.4$ and $\Omega_\Lambda \sim 0.6$. Therefore the universe should be presently dominated by a smooth component with effective negative pressure; this is in fact the most general requirement in order to explain the observed accelerated expansion. The most straightforward candidate for that is, of course, a "true" cosmological constant [27]. A plausible alternative that has recently received a great deal of attention is a dynamical vacuum energy given by a scalar field rolling down its potential: a cosmological scalar field, depending on its dynamics, can easily fulfill the condition of an equation of state $w_Q = p_Q/\rho_Q$ between -1 (which corresponds to the cosmological constant case) and 0 (that is the equation of state of matter). Since it is useful to have a short name for the rather long definition of this dynamical vacuum energy, we follow the literature in calling it briefly "quintessence"[28].

At the moment models with $\Omega_\Lambda \sim 0.6$ seem to be favored (see for example [29]). Ω_Λ is given by:

$$\Omega_\Lambda \equiv \frac{8\pi G \,\Lambda}{3H_0^2} \qquad (21)$$

where Λ is the cosmological constant, which appear in the most general form of the Einstein equation. The equation of state for Λ is $p = -\rho$ or equivalently $w = -1$. In order to have $\Omega_\Lambda \sim O(1)$, Λ has to be:

$$\Lambda \sim (2 \times 10^{-3} \text{ eV})^4. \qquad (22)$$

Being a constant there is no reason in Particle Physics why this constant should be so small and not receive corrections at the highest mass scale present in the theory. This constitutes the most severe hierarchy problem in Particle Physics and there are no hints in order to solve it.

If $\Lambda \neq 0$, in the Early Universe the density of energy and matter is dominant over the vacuum energy contribution, while the universe expands the average matter density decreases and at low redshifts the Λ term becomes important. At the end the universe starts inflating under the influence of the Λ term.

At present there are models based on the presence of Λ called ΛCDM models or ΛCHDM if we allow the presence of a small amount of HDM. Such models provide a good fit of the observed universe even if they need further studies and more data confirmations.

8. Neutrinos and the Matter-Antimatter Asymmetry in the Universe

It is an observational fact that the Universe (at least up to the level of galactic supercluster) exhibits a deep asymmetry between matter and antimatter. Unless this results from rather peculiar starting conditions of the Universe, one should envisage the existence of some dynamical mechanism to originate such asymmetry beginning from a symmetric situation in matter and antimatter.

The conditions which have to be fulfilled to obtain a baryogenesis have been individuated more than 30 years ago by Sackarov: we need processes which violate the baryon number B, the discrete symmetries C and CP which interchange baryons and antibaryons have also to be violated and, finally, the interactions responsible for the production of the baryonic asymmetry have to respect the so-called out-of-equilibrium condition so that, taking into account of the expansion of the Universe, one should prevent the Universe to have enough time to re-establish the originary symmetric situation after some asymmetry has been produced [30].

Now, in spite of the fact that B is conserved in the SM at all perturbative orders, B is actually violated at the quantum level since the baryonic current is anomalous. Indeed, only the combination B-L is conserved in the SM. This violation of B in the SM is extremely small today (and practically negligible for all practical purposes), however such B violations could play a major role in the early Universe when the temperature was of the order of the electroweak scale or higher [31].

The question then arises: is it possible to make use of the abovementioned B violations to have an efficient baryogenesis in the SM? The answer is negative. On one hand, the violation of CP in the SM turns out to be too small for a viable baryogenesis, on the other hand the electroweak phase transition is not sufficiently close to a first order phase transition to respect the out-of-equilibrium condition (the present lower bound on the Higgs mass clearly excludes the possibility of respecting the third baryogenesis condition).

In conclusion the need for dynamical mechanism for baryogenesis pushes us beyond the SM. There exist interesting examples of new physics which are suitable to obtain an efficient baryogenesis: in grand unified theories, in SUSY extensions of the SM and, interestingly enough, in models where neutrinos get a mass via a see-saw mechanism. We consider this latter possibility as one of the most attractive proposals to produce baryons. The idea is simple: it is not necessary to introduce new B violating interactions, but, rather, the violation of L which is present in the see-saw mechanism is enough to give rise to a lepton asymmetry, and, then, the presence of the B violating quantum effects of the ordinary electroweak interactions ensures that part of such L asymmetry is converted into a B asymmetry [32].

References

1. For an introduction to the DM problem, see, for instance:
   R. Kolb and S. Turner, in *The Early universe* (Addison-Wesley, New York, N.Y.) 1990;
   Dark Matter, ed. by M. Srednicki, (North-Holland, Amsterdam) 1989;
   J. Primack, D. Seckel and B. Sadoulet, *Annu. Rev. Nucl. Part. Sci.* **38**(1988) 751;
2. for a recent review see J. Primack, *astro-ph/0007187*;
3. S. Burles et al., *Phys. Rev. Lett.* **82**, 4176 (1999);
4. K. Freese, B.D. Fields, D.S. Graff, in *Proceedings of the MPA/ESO Workshop on the*

First Stars (Garching, Germany) August 4-6 1999, (astro-ph/0002058);
5. J.R. Bond, J.Centrella, A.S. Szalay, J.R.Wilson, in *Formation and evolution of Galaxies and Large Structures in the Universe*, ed. by J.Audouze, J. Tran Thanh Van (Reidel, Dordrecht) 1984, pp.87-99;
6. G.R. Blumenthal, J.R. Primack, in *Formation and evolution of Galaxies and Large Structures in the Universe*, ed. by J.Audouze, J. Tran Thanh Van (Reidel, Dordrecht) 1984, pp.163-83;
7. SuperK Coll., talk by H. Sobel at *Neutrino2000* Sudbury, Canada, June 16-21 2000;
8. G.L. Fogli, E. Lisi, A. Marrone, *Phys. Rev.* **D 64** (2001) 093005;
9. SuperK Coll., talk by Y. Suzuki, in *Neutrino2000* Sudbury, Canada, June 16-21 2000;
10. SNO Coll., Q.R. Ahmad et al., *Phys.Rev.Lett.* **87** 071301 (2001);
11. G.L. Fogli et al., *Phys.Rev.* **D 64** 093007 (2001); J. N. Bahcall, M. C. Gonzalez-Garcia, C. Pena-Garay, *JHEP* **0108** 014 (2001);
12. M. Apollonio et al., CHOOZ Collaboration, *Phys. Lett.* **B466**, 415 (1999);
13. G. Mills, in *Neutrino2000* Sudbury, Canada, June 16-21 2000;
14. V.M. Lobashev et al., *Phys. Lett.* **B460**, 227 (1999);
15. C. Weinheimer et al., *Phys. Lett.* **B460**, 219 (1999);
16. A. Osipowicz et al. (KATRIN Coll.), hep-ex/0109033;
17. H.V. Klapdor-Kleingrothaus et al., *Nucl. Phys. Proc. Suppl.* **100** 309 (2001);
18. S. Pascoli, S.T. Petcov and L. Wolfenstein, *Phys.Lett.* **B524** 319 (2002); S. Pascoli, S. T. Petcov, Talk at the Conference NANP'01, III International Conference on Non-Accelerator New Physics, Dubna (Russia), June 19-23 2001, *hep-ph/0111203*;
19. T. Yanagida, in *The Unified Theories and the Baryon Number in the Universe*, ed. O. Sawada and A. Sugamoto, (Tsukuba, Japan) 1979;
20. M. Gell-Mann, P. Ramond, R. Slansky, in *Supergravity*, ed. P. Van Nieuwenhuizen and D.Z. Freedman (1979);
21. U.L. Pen, U. Seljak, N. Turok, *Phys. Rev. Lett.* **79**, 1611 (1997);
22. A. Albrecht, R.A. Battye, J. Robinson, *Phys. Rev.* **D 59**, 023508 (1999);
23. S.Perlmutter et al., *Astrophys. J.* **517**, 565 (1999); S.Perlmutter et al., *Bull.Am.Astron.Soc.* **29**, 1351 (1997); S.Perlmutter et al., *Nature* **391**, 51 (1998); See also http://www-super-nova.LBL.gov/;
24. A.G.Riess et al., *Astron. J.* **116**, 1009 (1998); A.V.Filippenko and A.G.Riess, *Phys. Rept.* **307**, 31 (1998); A.V.Filippenko and A.G.Riess, in *Type Ia Supernovae: Theory and Cosmology*,Eds. Jens Niemeyer and Jim Truran, *astro-ph/9905049*; P.M.Garnavich et al, *Astrophys. J.* **501**, 74 (1998); B.Leibundgut, G.Contardo, P.Woudt, and J.Spyromilio, *astro-ph/9812042*; See also http://cfa-www.harvard.edu/cfa/oir/Research/supernova/HighZ.html;
25. J.G.Bartlett, A.Blanchard, M.Le Dour, M.Douspis, and D.Barbosa, *astro-ph/9804158*; G.Efstathiou, *astro-ph/9904356*; G.Efstathiou, S.L.Bridle, A.N.Lasenby, M.P.Hobson, and R.S.Ellis, *astro-ph/9812226*; C.Lineweaver, *Astrophys. J.* **505**, L69 (1998);
26. R.G.Carlberg, H.K.C.Yee, and E.Ellingson, *Astrophys. J.* **478**, 462 (1997); J.Carlstrom, 1999, *Physica Scripta*, in press.
27. See for example: S.M.Carroll, W.H.Press and E.L.Turner, *Annu. Rev. Astron. Astrophys.* **30**, 499 (1992);
28. R.R.Caldwell, R.Dave, and P.J.Steinhardt, *Phys. Rev. Lett.* **80**, 1582 (1998);
29. G.R. Blumenthal, S.M. Faber, J.R. Primack, M.J. Rees, *Nature* **311**, 517 (1984);
30. For a review on baryogenesis, see A. D. Dolgov, *Baryogenesis 30 Years Later*, hep-ph/9707419;
31. V. Kuzmin, V. Rubakov and M. Shaposhnikov, *Phys. Lett.* **B 155**, 36 (1985);
32. M. Fukugita and T. Yanagida, *Phys. Lett.* **B 174**, 45 (1986).

Recent Developments on Atmospheric Neutrinos
Chairperson: *F. Cervelli*

SIMULATION OF PARTICLE FLUXES IN THE EARTH'S VICINITY

VASSILI PLYASKIN*

Institute of Theoretical and Experimental Physics, 25 B.Cheremushkinskaya,117259,Moscow,Russia

The results of calculations of the fluxes of charged particles and atmospheric neutrinos are discussed. Results on charged particles are compared with the AMS measurements.

1. Introduction

In recent years much effort has been put into calculations of atmospheric showers[1]. A proper inclusion of the geomagnetic field effect on the initial flux of cosmic rays is the most important first step in these calculations. The approach when generation of the primary cosmic ray flux at the top of the atmosphere starts with emission of cosmic particles with an interstellar momentum spectra from a large (typically 10 Earth's radii) distance is recognized to be the most appropriate. The long distance from the emission point to the Earth is sufficient for a particle to endure the influence of the Earth's magnetic field to form an adequate topology of the flux in the Earth's vicinity. Calculations, following this direct tracing approach[2,3], describe well the AMS results on the fluxes of both primary and secondary particles at different magnetic latitudes. At the same time, the direct tracing calculation[3], while producing results on secondary lepton fluxes compatible with the AMS measurements, predicts atmospheric neutrino fluxes which are different from those obtained in previous calculations. These calculations use the back-tracing method when particles with interstellar rigidity spectra are isotropically emitted from a sphere situated at the the top of the atmosphere and are traced following the backward path of a particle with the same rigidity but the opposite charge. Only particles back-traced to a large distance (typically 10 Earth's radii) are considered as following an "allowed" trajectory and are included in the subsequent simulation of atmospheric showers. The back-tracing technique is considered as equivalent but computer time saving method of evaluation of the influence of the Earth's magnetic field on the primary cosmic ray flux.

As the comparison of results of calculations with data from atmospheric neutrino experiments[4] leads to some far reaching conclusions of fundamental importance, it is

---

*Present address EP division CERN, Geneva, Switzerland.

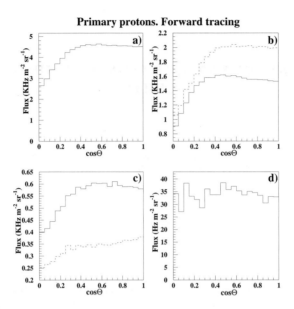

Fig. 1. Dependence of the average primary proton flux at 400 km on the proton incidence angle with respect to nadir. (a) All momenta. (b) Solid line:$p < 1.5$ GeV/c, dashed line:$1.5 < p < 5.0$ GeV/c. (c) Solid line:$5.0 < p < 10.0$ GeV/c, dashed line:$10.0 < p < 40.0$ GeV/c. (d) p> 40. GeV/c.

essential to confront the results of these calculations with all available experimental data on the fluxes of different particles in the vicinity of the Earth and on the ground. The data from the Alpha Magnetic Spectrometer (AMS)[5,6] on the fluxes of primary and secondary particles detected in the near Earth orbit covering most of the Earth's surface can be used as a gauge to assess the viability of different calculations. In the following, the results of calculations of particle fluxes at the Earth and the comparison of these results with AMS data are presented.

2. Description of the model

The present calculation is done in the framework of the model described in detail in [3]. The Earth is modeled to be a sphere of 6378.14 km radius. The Earth's magnetic field is calculated according to World Magnetic Field Model (WMM-2000)[8] with 6 degrees of spherical harmonics. The flux (Φ) of primary protons in the Solar system is parametrized on the basis of the AMS experimental data[5,7] by a power low in rigidity (R), $\Phi(R) \sim R^{-\gamma}$ with γ=2.79. At energies below several GeV the spectrum is corrected for the solar activity corresponding to the AMS flight period[9].

Primary cosmic particles are isotropically emitted from a sphere of 10 Earth's

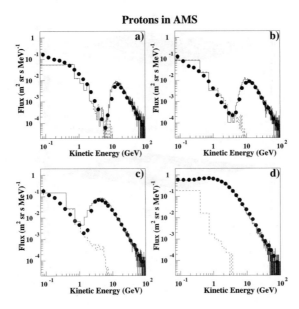

Fig. 2. Proton flux in AMS at different magnetic latitudes ($|\Theta_M|$). Solid line - flux from the top, dashed line - flux from the bottom. a) $|\Theta_M|$ <0.2 rad, b) 0.2< $|\Theta_M|$ <0.5 rad, c) 0.5< $|\Theta_M|$ <0.8 rad, c) 0.8< $|\Theta_M|$ <1.1 rad. The dots is the AMS measurement.

radii. Only particles with initial kinematics compatible with those reaching the Earth's atmosphere are traced the whole way until they interact with the atmosphere or the Earth itself. Charged secondaries from the interactions are traced until they go below 0.125 GeV/c or continue the journey beyond a sphere of 15 Earth's radii. The normalization of the fluxes is based on the interstellar flux measured by AMS[5,7]. The production of secondary pions and protons resulting from interactions of cosmic He with atmosphere is treated using the superposition approximation[10].

Under the influence of the Earth's magnetic field the primary flux becomes anisotropic for directions close to parallel to the Earth's surface. Fig.1 shows the angular distributions of primary proton fluxex of different rigidity at 400 km from the Earth's surface. The anisotropy is visible up to rigidity as high as 40 GV.

3. Particles detected by zenith facing AMS

To compare the results of calculations with the AMS measurements[5,6], the overall fluxes were restricted to the AMS acceptance and orbit during the zenith facing flight period.

Figure 2 shows the spectra of protons obtained in the simulation for different positions of AMS with respect to magnetic equator. The result of simulation cor-

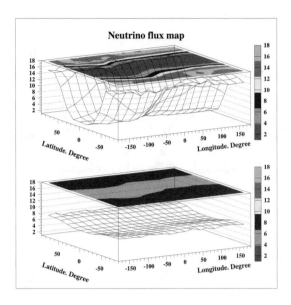

Fig. 3. The geographical map of the average atmospheric neutrino flux in KHz/m^{2}sr. Top - flux from the top. Bottom - flux from the bottom.

rectly reproduces the characteristic features of the data such as: the value of the cutoff at different magnetic latitudes; the presence of protons with under cutoff momenta at all magnetic latitudes; the equality of fluxes of under cutoff protons going to and from the Earth; the energy spectra of protons and their absolute flux. The simulated secondary positron and electron fluxes are similar to those measured by AMS[6] in terms of equality of fluxes of up and down going particles, dependence of energy spectra and e^+/e^- flux ratio on the magnetic latitude[3]. Although these positrons and electrons can not contribute to neutrino production they are produced in the same decays and are the "spectators" of atmospheric neutrino production. Consequently, the correct prediction of the fluxes of electrons and positrons can be considered as a proof of validity of the whole simulation procedure including neutrino flux predictions.

4. Neutrino flux

Although the flux of atmospheric neutrinos averaged over the entire Earth's surface is the same for neutrinos going into and out from the Earth, the actual neutrino flux for a given geographical location is by far different from the average one (Fig.3).

Moreover, the behaviour of calculated parameters describing the neutrino flux in the area e.g. within geographical coordinates of $35° \pm 12°$ N, $142° \pm 12°$ E close to the Kamioka site suggests systematic effects which may fake muon neutrino disappearance. The angular dependence of the ν_μ/ν_e ratio of upward going neutrino

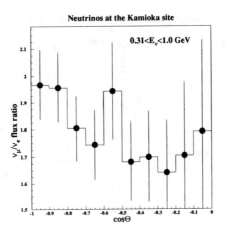

Fig. 4. Angular dependence of the ν_μ/ν_e ratio of upward going neutrino of energies $0.31 < E < 1.0$ GeV.

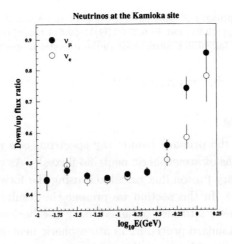

Fig. 5. Energy dependence of upward/downward neutrino flux ratio.

of energies $0.31 < E < 1.0$ GeV shown in Fig.4 is systematicaly lower than the value of 2.2 which is characteristic for the Earth surface average upward neutrino flux. Calculated for the same region, the energy dependence of the upward/downward neutrino flux ratio shown in Fig.5 also suggests a difference in the ratio for muon and electron neutrinos. It is clear, that a simulation with higher statistics is required

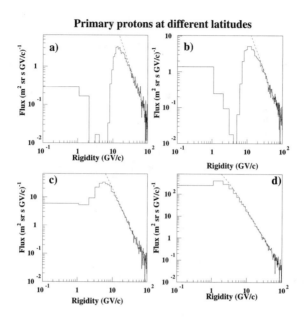

Fig. 6. Average flux of primary protons passing from the top a sphere at 400 km at different magnetic latitudes. a) $|\Theta_M| <0.253$ rad, b) $0.253< |\Theta_M| <0.524$ rad, c) $0.524< |\Theta_M| <0.848$ rad, d) $|\Theta_M| >0.848$ rad. The dashed line corresponds to the interstellar spectrum.

to make conclusion on the importance of these systematic effects.

5. Backward tracing

It is recognized, that the primary cosmic ray spectrum is a major source of uncertainty in calculations of atmospheric neutrino fluxes[11]. As it was demonstrated in section 2, the primary proton flux calculated using the forward tracing method fits well the AMS data. In this section we present the results on primary proton flux at different magnetic latitudes obtained using the backward tracing method commonly used as a standard procedure in atmospheric neutrino flux calculations. In this study the protons with interstellar rigidity spectra are isotropically emitted from a sphere situated at the the top of the atmosphere at 70 km from the Earth's surface and are traced following the backward path of a particle with the same rigidity but of the opposite charge. Only particles back-traced to a large distance (10 Earth's radii) are considered as following an "allowed" trajectory.

Fig.6 shows the calculated average flux of primary protons with "allowed" trajectories passing through a detection plane from the top hemisphere at different magnetic latitudes (Θ_M). The detection plane altitude is 400 km. Each of the four distributions in the figure corresponds to 1/4 of the total area of the Earth's surface.

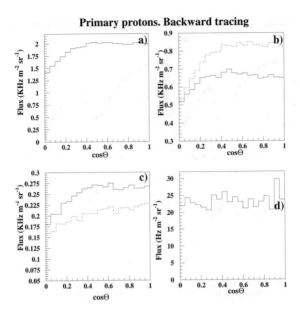

Fig. 7. Dependence of the average primary proton flux at 400 km on the proton incidence angle with respect to nadir. (a) All momenta. (b) Solid line:$p < 1.5$ GeV/c, dashed line:$1.5 < p < 5.0$ GeV/c. (c) Solid line:$5.0 < p < 10.0$ GeV/c, dashed line:$10.0 < p < 40.0$ GeV/c. (d) p> 40. GeV/c.

The dashed lines in the figure show the interstellar spectrum used in the calculations. Only in the regions close to the poles (Fig.6d) the calculated flux approaches the generated one in terms of intensity and the value of spectral index. For most of the Earth's surface the flux of primary protons predicted by the calculations is well below the interstellar one for all momenta above the magnetic cutoff value up to as high as 100GeV/c. As in the case of forward tracing (see Fig.1), there is an anisotropy of the primary flux for particles with rigidities below 40 GV, although the suppression of the flux for directions close to horizontal one is less pronounced. This difference has an important consequence for the production and decay of secondary particles in atmospheric showers.

A comparison of the result of back-tracing with AMS measurements[5,6] presented in Fig.8 shows that predictions of the flux of primary protons approaches the measured values only at high magnetic latitudes and is far below elsewhere.

6. Conclusions

The exact knowledge of primary cosmics flux value and topology near the Earth is a key point in calculations of the fluxes of secondary particles produced in atmospheric

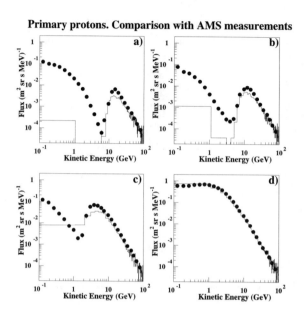

Fig. 8. Proton flux in AMS at different magnetic latitudes ($|\Theta_M|$). a) $|\Theta_M|$ <0.2 rad, b) 0.2< $|\Theta_M|$ <0.5 rad, c) 0.5< $|\Theta_M|$ <0.8 rad, c) 0.8< $|\Theta_M|$ <1.1 rad. The dots is the AMS measurement.

showers. The AMS data on particle fluxes provide a means to check the validity of predictions of different calculations.

Acknowledgments

I am grateful to the organizing committee, especially to R.Battiston, for an invitation to the workshop.

References

1. M.Honda et al, *Phys. Lett.* **B 248**, 193 (1990);
   M.Honda et al, *Phys. Rev.***D 52**, 4985 (1995);
   G.Barr,T.K.Gaisser and T.Stanev, *Phys. Rev.***D 39**, 3532 (1989);
   G.Barr,T.K.Gaisser and T.Stanev, *Phys. Rev.***D 38**, 85 (1989);
   V.Agraval et al, *Phys. Rev.***D 53**, 1314 (1996);
   L.V.Volkova, *Yad. Fiz.***31**, 1510 (1980) also in*Sov.J.Nucl.Phys.* **31**, 784 (1980).
   E.V.Bugaev and V.A.Naumov, *Phys. Lett.***B 232**, 391 (1989);
   A.V.Butkevich, L.G.Dedenko and I.M.Zheleznykh, *Yad.Fiz.***50**, 142 (1989) also in *Sov.J.Nucl.Phys.* **50**, 90 (1989).
   P.Lipari, *Astroparticle Physics* **1**, 195 (1993).
   G.Battistoni et al, *Astroparticle Physics* **12**, 315 (2000).
   P.Lipari,*Astroparticle Physics* **14**, 153 (2000).

M.Honda et al, Preprint arXiv:hep-ph/0103328 v2 (2001).
G.Battistoni,*Nucl.Phys.*B **100**, 101 (2001).
T.Gaisser, Preprint arXiv:hep-ph/0104327 v1 (2001).
G.Fiorentini et al, *Phys.Lett.*B **510**,173 (2001).

2. L.Derome et al,*Phys.Lett.*B **489**,1 (2000).
3. V.Plyaskin,*Phys.Lett.*B **516**,213 (2001); Preprint arXiv: hep-ph/0103286 v2 (2001).
4. Super-Kamiokande collaboration, Y.Fukuda et al, *Phys. Rev. Lett.***81**, 1562 (1998).
   Kamiokande collaboration, K.S.Hirata et al, *Phys. Lett.*B **205**, 416 (1988); *Phys. Lett.*B **289**, 146 (1992); Y.Fukuda et al,*Phys. Lett.*B **335**, 237 (1994).
   IMB collaboration, D.Casper et al, *Phys. Rev. Lett.***66**, 2561 (1991); R.Becker-Szendy et al,*Phys. Rev.*D **46**, 3720 (1992).
   Soudan collaboration, W.W.M.Allison et al,*Phys. Lett.*B **434**, 137 (1999).
   MACRO collaboration, *Phys. Lett.*B **434**, 451 (1998); Preprint hep-ex/0001044 submitted to Phys. Lett. B (2000).
5. AMS collaboration, *Phys. Lett.*B **472**, 215 (2000), *Phys. Lett.*B **490**, 27 (2000).
6. AMS collaboration, *Phys. Lett.*B **484**, 10 (2000).
7. AMS collaboration, *Phys. Lett.*B **494**, 193 (2000).
8. S.MacMillan et al, *Journal of Geomagnetism and Geoelectricity***49**, 229 (1997); J.M.Quinn et al, *Journal of Geomagnetism and Geoelectricity***49**, 245 (1997);
9. W.R.Webber and. M.S.Potgieter, *ApJ.***344**, 779 (1989).
10. H.R.Schmidt,J.Schukraft, *J.Phys.*G **19**, 1705 (1993).
11. P.Lipari,*Nucl.Phys.*B **100**, 136 (2001).

CALCULATION OF SECONDARY PARTICLES IN ATMOSPHERE AND HADRONIC INTERACTIONS

G. BATTISTONI

INFN and Università di Milano, Dipartimento di Fisica, Milano, 20133, Italy.

A. FERRARI

CERN, Geneva 23, Switzerland, on leave of absence from INFN, Milano.

P.R. SALA

ETH, Zurich, Switzerland, on leave of absence from INFN, Milano.

Calculation of secondary particles produced by the interaction of cosmic rays with the nuclei of Earth's atmosphere pose important requirements to particle production models. Here we summarize the important features of hadronic simulations, stressing the importance of the so called "microscopic" approach, making explicit reference to the case of the FLUKA code. Benchmarks are also presented

Keywords: Cosmic Rays; MonteCarlo; Hadronic Interactions.

1. Introduction

Reliable calculations of flux of secondary particles in atmosphere, produced by the interaction of primary cosmic rays, are essential for the correct interpretation of the large amount of experimental data produced by experiments in the field of astroparticle physics. The increasing accuracy of modern experiments demands also an improved quality of the calculation tools. Different ingredients are required to produce a useful calculation model. Essentially they can be reduced to three important classes: the primary cosmic ray spectrum, the modelization of the environment (atmosphere, geomagnetic field, etc.) and a model of particle production in the hadronic shower following the collisions of primary c.r.'s with the atmosphere nuclei. The uncertainty on the primary spectra is dominated by the systematics of the experiments devoted to their measurement: in the light of recent measurements by AMS[1] and BESS[2], such an uncertainty is about ±5% below 100 GeV/nucleon, increasing to ±10% at 10 TeV/nucleon. As far as the environmental description is concerned, the large amount of geophysical data now available allows, in principle, to achieve a high level of accuracy. On the other hand, the knowledge of the features of particle production in hadronic interactions is still affected by important uncertainties ($\gtrsim$ ±15%). Since we have not yet a calculable theory for the non-perturbative QCD regime, we remain with many different attempts of building interaction models, which are tuned by comparison to existing experimental data. Sometimes, these models can give satisfactory outputs only in restricted fields of applications. In this work we discuss in some more detail the situation of these attempts to describe hadronic interactions, trying to evidentiate the advantages of the so called "microscopic" models, *i.e.* those which try to embed as much as possible of the current theoretical ideas in terms of elementary constituents and of their fundamental interactions. We shall make explicit reference to the set of models contained in the FLUKA MonteCarlo code[3].

2. Requirements for Interaction Models

Cosmic ray physics is particularly demanding from the point of view of particle production models, since it is in general necessary to consider a wide range of primary energy and projectiles. This is quite a different situation with respect to the standard case of particle physics, where, in general, almost mono-energetic beams are considered. In addition, while in particle physics the attention is mainly devoted to energy deposition, here instead the details of single interactions are fundamental to obtain a flux prediction. As previously stated, it is not possible to rely on a unique model capable of giving the same quality of results at all energies, and for all kinematics regimes. Therefore particular attention is necessary in order to assure the right continuity across the transition region between the different regimes.

Last, but not least, there is the necessity of dealing with different nuclear species. So far, this problem has been mainly solved by recurring to the so called "superposition" model, where a nucleus of mass number A and energy E_0 is considered to be equivalent to A nucleons, each one having energy E_0/A. The question if this is a totally acceptable approximation is still an argument of discussion.

3. Different approaches: parametrized vs "microscopic" models

In building a suitable model for particle production, we can identify two main different attitudes: parametrized codes and theoretically inspired simulations. In the first case, one relies upon analytical formulas, parametrizing experimental data, which possibly try to exploit some general phenomenological features of particle production. An example of these is Feymann scaling, which is known to be a rather good, although not exact, approximation, especially in the forward ("fragmentation") kinematic region which is dominant in secondary production by cosmic rays. This property can be expressed as follows. The number of pions of energy E_π produced in an interaction by a primary proton of energy E_0 is well represented expression:

$$\frac{dn_\pi}{dE_\pi}(E_\pi, E_0) \simeq \frac{1}{E_\pi} F\left(\frac{E_\pi}{E_0}\right) \quad (1)$$

where the function $F(x)$, $x = E_\pi/E_0 \sim x_{Feynman}$, is approximately independent from the primary energy E_0, and decreases monotonically from a finite value for $x \to 0$, to zero for $x \to 1$. The shape of the curve just needs few parameters that can be extrated from experimental data sets. This kind of approach is at the basis of the work which has been performed by the Bartol group for many years, producing many valuable results, and in particular the prediction for atmospheric neutrino fluxes[4]. For this purpose, they constructed the TARGET numerical model, a module which can be easily inserted in any cascade program[a]. TARGET considers hadron interactions on light nuclei, like Oxygen and Nitrogen, subdividing the available energy between leading nucleons and other produced hadrons on the basis of an assumed elasticity function. Pions, kaons, etc. are then produced according to parametric formulas reproducing the scaling properties described above. Experimental data at different energies fix the parameters and guide the evolution of multiplicities as a function of energy. At low and intermediate energies, resonance production is considered. Care has been taken to assure event by event energy conservation. The advantage of this kind of approach, mainly used in the framework of a 1-dimensional description, is that it can lead to the comprehension of some important and general properties of particle production in terms of analytical expressions. The price to pay is the lack of generality and the spoiling of correlations among reaction products.

The second line of approach is instead the use of models which try to describe interactions in terms of the properties of elementary constituents. In principle one would like to derive all features of "soft" interactions (low-p_T interactions) from the QCD Lagrangian, as it is done for hard processes. Unfortunately the large value taken by the running coupling constant prevents the use of perturbation theory. Indeed, in QCD, the color field acting among quarks is carried by the vector bosons of the strong interaction, the gluons, which are "colored" themselves. Therefore the characteristic feature of gluons (and QCD) is their strong self-interaction. If we imagine that quarks are held together by color lines of force, the gluon-gluon interaction will pull them together into the form of a tube or a string. Since quarks are confined, the energy required to "stretch" such a string is increasingly large until it suffices to materialize a quark-antiquark couple from the vacuum and the string breaks into two shorter ones, with still quarks at both ends. Therefore it is not unnatural that because of quark confinement, theories based on interacting strings emerged as a powerful tool in understanding QCD at the soft hadronic scale (the non-perturbative regime). Different implementations of this idea exist, having obtained remarkable success in describing the features of hadronic interactions. In the following section we shall concentrate on the example of the FLUKA MonteCarlo code.

We consider the microscopic approach as more fundamental and reliable than parametrizations, for many reasons, since each step has sound physical basis. The performances are optimized comparing with particle production data at single interaction level. The final predictions are obtained with a minimal set of free parameters, fixed for all energies and target/projectile combinations. Results in complex

---

[a] an upgrade of the TARGET model, also in view of 3-D applications, has been presented in [5]

cases as well as scaling laws and properties come out naturally from the underlying physical models. The basic conservation laws are fulfilled "a priori". A microscopic model can reach a very high level of detail, at least in principle, and therefore is a good choice when aiming at precision calculations. The price to pay is the loss of simplicity and flexibility: there are no more simple analytical guidelines which allow to understand the basic properties. Furthermore microscopic codes are more demanding than parametrizations in terms of computing power.

In summary, in order to reach a deeper understanding and reliability of predictions, microscopic models are necessary. Parametrized models (if parametrizations are performed at the level of single interactions) are instead useful as a first, fast and flexible approach. Instead, models tuned on "integral data", like calorimeter resolutions, thick target yields etc., can be very inaccurate at the level of single interactions, as shown in ref. [6] for the case of *GEANT-GHEISHA*: such a model cannot be used to obtain a reliable calculation of particle fluxes.

4. The FLUKA model

The modern FLUKA[3] is an interaction and transport MonteCarlo code able to treat with a high degree of detail the following problems:

- Hadron-hadron and hadron-nucleus interactions 0-100 TeV
- Electromagnetic and μ interactions 1 keV-100 TeV
- Charged particle transport - ionization energy loss
- Neutron multigroup transport and interactions 0-20 MeV
- Nucleus-nucleus and hadron-nucleus interactions 0-10000 TeV/n: *under development*

Here we shall review the two hadronic models which are used inside FLUKA to describe nonelastic interactions:

- The "low-intermediate" energy one, PEANUT, which covers the energy range up to 5 GeV
- The high energy one which can be used up to several tens of TeV, based on the color strings concepts sketched in the previous section.

The nuclear physics embedded in the two models is very much the same. The main differences are a coarser nuclear description (and no preequilibrium stage) and the Gribov-Glauber cascade for the high energy one.

4.1. *The PEANUT Model*

Hadron-nucleus non-elastic interactions are often described in the framework of the IntraNuclear Cascade (INC) models. This kind of model was developed at the very beginning of the history of energetic nuclear interaction modelling, but it is still valid and in some energy range it is the only available choice. Classical INC codes were based on a more or less accurate treatment of hadron multiple collision processes in nuclei, the target being assumed to be a cold Fermi gas of nucleons in their potential well. The hadron-nucleon cross sections used in the calculations are free hadron–nucleon cross sections. Usually, the only quantum mechanical concept incorporated was the Pauli principle. Possible hadrons were often limited to pions and nucleons, pions being also produce or absorbed via isobar (mainly Δ_{33}) formation, decay, and capture. Most of the historical weaknesses of INC codes have been mitigated or even completely solved in some of the most recent developments [3,7], thanks to the inclusion of a so called "preequilibrium" stage, and to further quantistic effects including coherence and multibody effects.

All these improvements are considered in the PEANUT (PreEquilibrium Approach to NUclear Thermalization) model of FLUKA. Here the reaction mechanism is modelled in by explicit intranuclear cascade smoothly joined to statistical (exciton) preequilibrium emission [8] and followed by evaporation (or fission or Fermi break-up) and gamma deexcitation. In both stages, INC and exciton, the nucleus is modelled as a sphere with density given by a symmetrized Woods-Saxon [9] shape with parameters according to the droplet model [10] for A>16, and by a harmonic oscillator shell model for light isotopes (see [11]). The effects of the nuclear and Coulomb potentials outside the nuclear boundary are included. Proton and neutron densities are generally different. Binding Energies are obtained from mass tables. Relativistic kinematics is applied at all stages, with accurate conservation of energy and momentum including those of the residual nucleus. Further details and validations can be found in [3].

For energies in excess of few hundreds MeV the inelastic channels (pion production channels) start to play a major role. The isobar model easily accommodates multiple pion production, for example allowing the presence of more than one resonance in the intermediate state (double pion production opens already at 600 MeV in nucleon-nucleon reactions, and at about 350 MeV in pion-nucleon ones). Resonances which appear in the intermediate states can be treated as real particles, that is, they can be transported and then transformed into secondaries according to their lifetimes and decay branching ratios.

4.2. *The Dual Parton Model for high energy*

A theory of interacting strings can be managed by means of the Reggeon-Pomeron calculus in the framework of perturbative Reggeon Field Theory[12], an expansion already developed before the establishment of QCD. Regge theory makes use explicitly of the constraints of analyticity and duality. On the basis of these concepts, calculable models can be constructed and one of the most successful attempts in this field is the so called "Dual Parton Model" (DPM), originally developed in Orsay in 1979 [13]. It provides the theoretical framework to describe hadron-nucleon interaction from several GeV onwards. In DPM a hadron is a low-lying excitation of an open string with quarks, antiquarks or diquarks sitting at its ends. In particular mesons are described as strings with their valence quark and antiquark at the ends. (Anti)baryons are treated like open strings with a (anti)quark and a (anti)diquark at the ends, made up with their valence quarks.

At sufficiently high energies, the leading term in high energy scattering corresponds to a "Pomeron" ($I\!P$) exchange (a closed string exchange with the quantum numbers of vacuum), which has a cylinder topology. By means of the optical theorem, connecting the forward elastic scattering amplitude to the total inelastic cross section, it can be shown that from the Pomeron topology it follows that two hadronic chains are left as the sources of particle production (unitarity cut of the Pomeron). While the partons (quarks or diquarks) out of which chains are stretched carry a net color, the chains themselves are built in such a way to carry no net color, or to be more exact to constitute color singlets like all naturally occuring hadrons. In practice, as a consequence of color exchange in the interaction, each colliding hadron splits into two colored system, one carrying color charge c and the other $\bar{c}$. These two systems carry together the whole momentum of the hadron. The system with color charge c ($\bar{c}$) of one hadron combines with the system of complementary color of the other hadron, in such a way to form two color neutral chains. These chains appear as two back-to-back jets in their own centre-of-mass systems. The exact way of building up these chains depends on the nature of the projectile-target combination (baryon-baryon, meson-baryon, antibaryon-baryon, meson-meson). Let us take as example the case of nucleon-nucleon (baryon-baryon) scattering. In this case, indicating with q_p^v the valence quarks of the projectile, and with q_t^v those of the target, and assuming that the quarks sitting at one end of the baryon strings carry momentum fraction x_p^v and x_t^v respectively, the resulting chains are $q_t^v - q_p^v q_p^v$ and $q_p^v - q_t^v q_t^v$, as shown in fig. 1.

Energy and momentum in the centre-of-mass system of the collision, as well as the invariant mass squared of the two chains, can be obtained from:

$$E^*_{ch1} \approx \frac{\sqrt{s}}{2}(1 - x_p^v + x_t^v)$$

$$E^*_{ch2} \approx \frac{\sqrt{s}}{2}(1 - x_t^v + x_p^v)$$

$$p^*_{ch1} \approx \frac{\sqrt{s}}{2}(1 - x_p^v - x_t^v) = -p^*_{ch2} \quad (2)$$

$$s_{ch1} \approx s(1 - x_p^v)x_t^v$$

$$s_{ch2} \approx s(1 - x_t^v)x_p^v$$

The single Pomeron exchange diagram is the dominant contribution, however higher order contributions with multi-Pomeron exchanges become important at energies in excess of 1 TeV in the laboratory. They correspond to more complicated topologies, and DPM provides a way for evaluating the weight of each, keeping into account the unitarity constraint. Every extra Pomeron exchanged gives rise to two extra chains which are built using two $q\bar{q}$ couples excited from the projectile and target hadron sea respectively. The inclusion of these higher order diagrams is

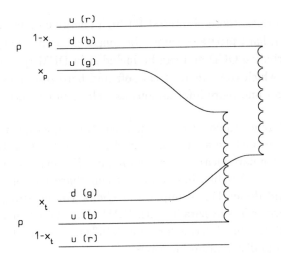

Fig. 1. Leading two-chain diagram in DPM for $p-p$ scattering. The color (red, blue, and green) and quark combination shown in the figure is just one of the allowed possibilities.

usually referred to as *multiple soft collisions*.

Two more ingredients are required to completely settle the problem. The former is the momentum distribution for the x variables of valence and sea quarks. Despite the exact form of the momentum distribution function, $P(x_1,..,x_n)$, is not known, general considerations based on Regge arguments allow to predict the asymptotic behavior of this distribution whenever each of its arguments goes to zero. The behavior turns out to be singular in all cases, but for the diquarks. A reasonable assumption, always made in practice, is therefore to approximate the true unknown distribution function with the product of all these asymptotic behaviors, treating all the rest as a normalization constant.

The latter ingredient is a hadronization model, which must take care of transforming each chain into a sequence of physical hadrons, stable ones or resonances. The basic assumption is that of *chain universality*, which assumes that once the chain ends and the invariant mass of the chain are given, the hadronization properties are the same regardless of the physical process which originated the chain. Therefore the knowledge coming from hard processes and e^+e^- collisions about hadronization can be used to fulfill this task. There are many more or less phenomenological models which have been developed to describe hadronization (examples can be found in [14,15]). In principle hadronization properties too can be derived from Regge formalism [16].

It is possible to extend DPM to hadron-nucleus collisions too [13], making use of the so called Glauber-Gribov approach. Furthermore DPM provides a theoretical framework for describing hadron diffractive scattering both in hadron-hadron and hadron-nucleus collisions. General informations on diffraction in DPM can be found in [17] and details as well as practical implementations in the DPM framework in [18].

At very high energies, those of interest for high energy cosmic ray studies (10–10^5 TeV in the lab), hard processes cannot be longer ignored. They are calculable by means of perturbative QCD and can be included in DPM through proper unitarization schemes which consistently treat soft and hard processes together. The interested reader can find more informations as well as practical implementations and results in [13,19].

DPM exhibited remarkable successes in predicting experimental observables. The quoted references include a vast amount of material showing the capabilities of the model when compared with experimental data. However, it must be stressed that other models are available, but most of them share an approach based on string formation and decay. For example, the *Quark Gluon String Model* [20] has been developed more or less in parallel with DPM. This model shares most of the basic features of DPM, while differing for some details in the way chains are created and in the momentum distribution functions.

5. Benchmarks of the FLUKA Model

The predictions of FLUKA have been checked with a large set of experimental data collected in accelerator experiments. Here we shall limit ourselves to show only a few examples, among the most important in view of the application of the code to cosmic ray applications.

Two sets of data are of particular relevance to check the quality of a model to be used for the calculation of atmospheric neutrino fluxes. These concern p-Be collisions and are reported in fig. 2: in ref.[21] the central rapidity region has been mainly explored, while in ref.[22] the forward region has been investigated. In both cases the agreement of FLUKA predictions is quite good.

Measurements of $\pi^\pm$ and $K^\pm$ production rates by 400 GeV/c protons on Be targets were performed by Atherton et al. [23] for secondary particle momenta above 60 GeV/c and up to 500 MeV/c of transverse momentum. Recently the NA56/SPY (Secondary Particle Yields) experiment [24] was devoted to directly measure these yields in the momentum region below 60 GeV/c. The SPY experiment measured the production at different angles θ and momenta $P \leq 135$ GeV/c down to 7 GeV/c for pions, kaons, protons and their antiparticles, using a 450 GeV/c proton beam impinging on Be targets. These data were extremely valuable to improve the hadronization model of FLUKA so to arrive at the present version. FLUKA is in agreement with the Atherton and the SPY measurements at the level of $\sim$ 20% in the whole momentum range of all secondaries, with the exception of a few points mostly for negative kaons. The case of pions is reported in fig.3. Also the θ dependence of the measured yields is reasonably described by FLUKA. The measured $K^\pm/\pi^\pm$ ratios are reproduced to better than 20% below 120 GeV/c.

This example was of particular relevance, since other attempts, using for instance the hadronic interfaces of *GEANT* (and in particular *GEANT-GHEISHA*) yielded a much worser agreement, as shown in ref.[25].

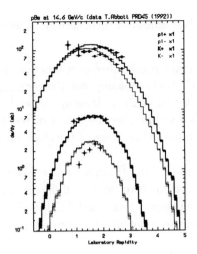

Fig. 2. Rapidity distribution of $pi^{+/-}$ and $K^{+/-}$ for 14.6 GeV/c protons on Be (left, data from ref. [21]), and X_{lab} distribution for $pi^{+/-}$ for 24 GeV/c protons on Be (right, symbols extrapolated from the double differential cross section reported in ref. [22]). Histograms are simulation results.

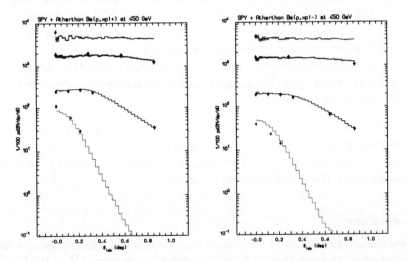

Fig. 3. Double differential cross section for pi^+ (left) and π^- (right) production for 450 GeV/c protons on a 10 cm thick Be target (data from ref. [23] and [24]). Data are given as a function of θ_{lab} and for different momentum bins. From top to bottom: 7, 15, 40 and 135 GeV/c, scaled respectively by a factor of 19683, 2187, 81 and 9. Histograms are simulation results.

6. Example of calculations of particles in atmosphere

In the last years, the FLUKA interaction models has been used to produce new predictions for the atmospheric neutrino fluxes within a full 3D calculation[27]. These fluxes are now also considered by the Super–Kamiokande experiment[28]. In the last two years a considerable amount of work has been devoted to cross check the validity of the calculation model. As far as the FLUKA approach is concerned, at least two remarkable results can be quoted:

(1) The reproduction of the features of primary proton flux as a function of geomagnetic latitude as measured by AMS[29], thus showing that the geomagnetic effects and the overall geometrical description of the 3-D setup are well under control. In addition, the same work shows that also the fluxes of secondary e^+e^- measured at high altitude (eventually the last stage of the chain decay of produced mesons) are reproduced.
(2) The good reproduction of the data on muons in atmosphere as measured by the CAPRICE experiment[30], both at ground level and at different floating altitudes[31], when starting from the same primary flux (Bartol fit) used to generate atmospheric neutrinos. See the quoted reference for relevant plots and numbers.

The fluxes of atmospheric muons are strictly related to the neutrino ones, because almost all ν's are produced either in association, with, or in the decay of $\mu^{\pm}$. Therefore it is possible to conclude that, for that choice of primary spectrum, the ν fluxes predicted by FLUKA are probably in the right range. To a large extent the agreement between the original HKKM[32] and Bartol[4] calculations of the ν fluxes, despite they started from different estimates of the primary flux and different hadronic interaction models, is not casual, but the result of the μ constraint. Furthermore, the agreement exhibited by the FLUKA simulation for muons of both charges gives confidence on the predictions of FLUKA for the parent mesons of muons (mostly pions).

The shower simulations in atmosphere have been compared also to the most recent hadron spectra at different latitudes and altitudes, obtaining remarkable agreement. As an example, in fig. 4 we compare MonteCarlo results to the hadron flux measured with the KASKADE experiment[33].

7. Conclusions

The phenomenological study of hadronic interactions is still a fundamental issue for astroparticle physics. Now that, at least for primary energy lower than 100 GeV, the uncertainties on primary spectrum have been substantially reduced by the quality of new experiments like AMS[1] and BESS[2], the need for a better quality model of hadronic interactions is even more necessary if accuracy of predictions has to be pursued. For this goal the "microscopic" codes are mostly recommended, thanks to their predictive power in a very large kinematic region, constrained by a limited number of parameters. Uncertainties on the modelling of hadronic interactions will remain a fundamental issue, and probably only new data, if experimental systematics can be kept under reasonable control, will help model builders. The HARP experiment[26] at CERN is aiming at this goal. This kind of activity is beneficial not only for particle physics and astrophysics, but also for applied science, since

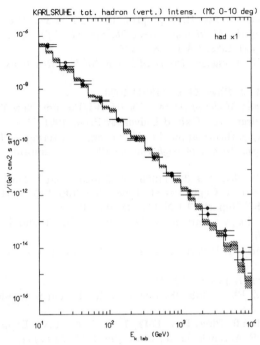

Fig. 4. Hadron flux measured with the KASKADE experiment [33]. Histogram is simulation result.

these calculations are necessary to understand radiation fluxes in the Earth's atmosphere, and this is of great interest for civil aviation and for the design of satellite activities[34]. The FLUKA MonteCarlo model is already being used for this purpose: doses to commercial flight are the subject of a work in progress, together with the development of a specific model for heavy ion transport and interaction: this will be of the utmost importance for dose and damage calculation in space aircrafts.

References

1. J. Alcaraz et al. (the AMS Collaboration), Phys.Lett. B 472, 215 (1999).
2. Bess Collaboration (T. Sanuki et al.), Ap.J. 545, 1135 (2000). [astro ph/0002481]
3. A. Fassò, A. Ferrari, J. Ranft, P.R. Sala, *FLUKA: Status and Prospective for Hadronic Applications*, invited talk in Proceedings of the MonteCarlo 2000 Conference, Lisbon, October 23–26 2000, A. Kling, F. Barão, M. Nakagawa, L. Távora, P. Vaz eds., Springer-Verlag Berlin, p. 955-960 (2001). Other references and documentation on FLUKA are available at http://www.cern.ch/fluka.
4. V. Agrawal et al., Phys. Rev. **D53** (1996) 1314.
5. T.K. Gaisser, proc. of Taup2001, Sep 8-12 2001, Assergi (Italy); R. Engel, et al. Proc. of the 27th ICRC (Hamburg, 2001), Session HE2.02.
6. F. Carminati and I. Gonzales Caballero, ALICE internal note ALICE-INT-2001-041, CERN, (2001). Available at
   http://edmsoraweb.cern.ch:8001/cedar/doc.info?document_id=331045&version=1
7. S.G Mashnik and S.A. Smolyansky, *The cascade-exciton approach to nuclear reactions:*

foundation and achievements, JINR preprint 1994: E2-94-353; R.E. Prael and M. Bozoian, *Adaptation of the Multistage Preequilibrium model for the MonteCarlo Method (I)*, Los Alamos report 1988: LA-UR-88-3238.
8. E. Gadioli and P. E. Hodgson, *Pre-equilibrium Nuclear Reactions*, Clarendon Press, Oxford, (1992).
9. M.E. Grypeos et al., J. Phys. **G 17** (1991) 1093.
10. W.D. Myers, *Droplet Model of Atomic Nuclei*, IFI/Plenum, New York, 1977.
11. L.R.B. Elton, *Nuclear sizes*, Oxford University Press, 1961.
12. For a review of Regge theory applied to high energy scattering see P.D.B. Collins, *An Introduction to Regge Theory & High Energy Physics*, (Cambridge University Press, Cambridge 1977).
13. A. Capella et al., Z. Phys. **C3**, 329 (1980); A. Capella, and J. Tran Thanh Van, Phys. Lett. **B93**, 146 (1980); A. Capella et al., Phys. Rep. **236**, 225 (1994).
14. T. Sjostrand, CERN Report CERN-TH 6488/92 (1992).
15. S. Ritter, *Comput. Phys. Commun.* 31, 393 (1984); J. Ranft, and S. Ritter, Acta Phys. Pol. **B11** 259 (1980).
16. A.B. Kaidalov, and O.I. Piskunova, Z. Phys. **C30**, 141 (1986); A. Capella et al., Z. Phys. **C70**, 507 (1996).
17. K. Goulianos, Phys. Rep. **101**, 169 (1983).
18. S. Roesler et al., Z. Phys. **C59**, 481 (1993); J. Ranft, and S. Roesler, Z. Phys. **C62**, 329 (1994).
19. K. Hahn, and J. Ranft, Phys. Rev. **D41**, 1463 (1990); F.W. Bopp et al., Phys. Rev **D49**, 3236 (1994); P. Aurenche et al., Phys. Rev **D45**, 92 (1992).
20. A. Kaidalov, Phys. Lett. **B117**, 459 (1982); A. Kaidalov, and K.A. Ter-Martirosyan, *Phys. Lett.* B117, 247 (1982).
21. T. Abbott et al Phys. Rev. **D45(11)**, 3906 (1992)
22. T. Eichten et al. Nucl. Phys. **B44**, 333 (1972).
23. H.W. Atherton, CERN 80-07 (1980).
24. G. Ambrosini et al., Phys. Lett. **B425** 208 (1988).
25. G. Collazuol et al., Nucl. Instr. & Meth. **A 449** 609 (2000).
26. See http://harp.web.cern.ch/harp/
27. G. Battistoni et al., Astrop. Phys. **12** (2000) 315. FLUKA flux tables are available at http://www.mi.infn.it/~battist/neutrino.html.
28. M. Shiozawa, for the Super–Kamiokande collaboration, these proceedings.
29. P. Zuccon et al., these proceedings. Also: proc. of Taup2001, Sep. 8-12 2001, Assergi, Italy. Also: astro-ph/0111111, submitted to Phys. Lett. B.
30. M. Boezio et al., Phys. Rev. **D62** (2000) 032007.
31. G. Battistoni et al., hep-ph/0197241, to be published in Astroparticle Phys.
32. M. Honda et al., Phys. Rev. **D52** (1995) 4985.
33. H.H Mielke et al., *Cosmic ray hadron flux at sea level up to 15 TeV*, Journ. Phys. G 1994: 20; 637-650, and H. Kornmayer et al, *High-energy cosmic-ray neutrons at sea level*, Journ. Phys. G 1995: 21; 439-450.
34. S. Roesler et al., *Monte Carlo Calculation of the Radiation field at Aircraft Altitudes*, SLAC-PUB-8968 (2001); A. Ferrari, M. Pelliccioni and T. Rancati, *A Method Applicable to Effective Dose Rate Estimates for Aircrew Dosimetry*, Radiation Protection Dosimetry, **96**, 219-222 (2001).

MASSIVE NEUTRINOS AND THEORETICAL DEVELOPMENTS

ALESSANDRO STRUMIA[*]

Theoretical Physics Division, CERN, CH-1211 Geneva 23, Switzerland

FRANCESCO VISSANI

Theory Group, Laboratori Nazionali del Gran Sasso, S.S. 17bis, km 18+910, Assergi (AQ), I-67010, Italy

This presentation includes recent developments on massive neutrinos and introductory material. Most discussion stays close to the data (=phenomenology), but it is integrated with some more formal (=theoretical) remarks. The importance of oscillations for supernova neutrinos is emphasized, and the peculiar role of ν_e events in this connection is discussed. A new bound on neutrino mass scale, based on double beta decay bound, is obtained.

Keywords: Neutrino mass and mixing; neutrino oscillations; neutrinoless $\beta\beta$ decay.

1. Massive Neutrinos

1.1. The Parameters

The main case under discussion is 3 neutrinos with Majorana masses:

$$\mathcal{L} = \frac{1}{2} \nu_\ell \, \mathbf{M}_{\ell\ell'} \, \nu_{\ell'} + \text{h.c.} \qquad \ell, \ell' = e, \mu, \tau$$

It is useful to decompose the mass matrix as follows:

$$\mathbf{M}_{\ell\ell'} = U^*_{\ell j} \cdot m_j \cdot e^{i\xi_j} \cdot U^*_{\ell' j}$$

$\mathbf{M}_{\ell\ell'}$ has 9 physical parameters. It might be possible to measure all but 1 of them:

- the 3 angles θ_{ij} and the phase δ from oscillations.[a]
- $\Delta m^2_{ij} = m_i^2 - m_j^2$ from oscillations (2 independent parameters).
- $|\mathbf{M}_{ee}|$ from neutrinoless double beta decay ($0\nu 2\beta$).
- $m_{\nu_e} = (\mathbf{M}^\dagger \mathbf{M})_{ee}$ from β decay.

They will be discussed in some length below. Today we know 3 of them (Δm^2_{32}, θ_{23} and θ_{12}), and have limits on other 4 (Δm^2_{21}, $\mathbf{M}_{ee}$, m_{ν_e} and θ_{13}). Note that the

---

[*]On leave from Dipartimento di Fisica dell'Università di Pisa and INFN.
[a]These are the parameters of the decomposition of the neutrino mixing matrix: $U_{e3} = \sin\theta_{13}\exp(-i\delta)$, $U_{e2}/U_{e1} = \tan\theta_{12}$, $U_{\mu 3}/U_{\tau 3} = \tan\theta_{23}$, where $0 \le \theta_{ij} < 90°$. The other elements are fixed by unitarity.

parameters ξ_j (Majorana phases) are not accessible by oscillations or by β decay; we could learn something on them from $\mathbf{M}_{ee}$ (from $0\nu2\beta$).[b]

1.2. Basics of Oscillations

This phenomenon[1] is due to the mismatch among states at production, propagation and detection. Indeed, ultra-relativistic neutrinos ($p \gg m_i$) of flavor ℓ produced by weak interactions are the superposition $|\nu_\ell\rangle = U_{\ell i}^* |\nu_i\rangle$. During propagation in vacuum, the state $|\nu_2\rangle$ acquires a phase with respect to $|\nu_1\rangle$ equal to $\exp[-i\Delta m_{21}^2 t/(2p)]$ (similarly for $|\nu_3\rangle$). Therefore, after propagation there is some overlap with neutrinos with different flavor $|\nu_{\ell'}\rangle$ (appearance) and at the same time, there is a reduced overlap with the original flavor $|\nu_\ell\rangle$ (disappearance). The amplitude of propagation can be rewritten as $\exp(-iH_{\text{eff}} t)$, where $H_{\text{eff}} = \mathbf{M}^\dagger \mathbf{M}/2E$ since $p \sim E$. Note that oscillations are not affected by the overall mass scale and by Majorana phases[2]. 2 flavor formulæ are widely used: 2 examples (pertinent to appearance and disappearance) are $P_{\mu \to \tau} = \cos^4 \theta_{13} \sin^2 2\theta_{23} \sin^2 \varphi$ and $P_{e \to e} = 1 - \sin^2 2\theta_{13} \sin^2 \varphi$ (probabilities are obtained on squaring the amplitudes). In both formulæ, the phase of oscillation is $\varphi = 1.267 \Delta m^2/\text{eV}^2 \, L/\text{m}, \text{MeV}/E$ (since $L \approx t$) where Δm^2 is the relevant mass splitting. We recall that at present there are evidences of 2 flavor oscillations, not of 3 flavor oscillation. Indeed, CHOOZ results[3] imply that there is no evident disappearance of electron neutrinos for the L/E values relevant to the interpretation. This boils down in a bound on the angle θ_{13}, and in an (approximate) decoupling of the oscillations with big and small Δm^2, that corresponds to atmospheric and solar neutrino oscillations respectively (we do not include in the discussion a very interesting LSND result that could be interpreted as oscillations but would require a fourth neutrino or something else[4]).

When electron neutrinos propagate in a medium, they acquire an additional phase of scattering due to charged current interactions with electrons. This results in another contribution to H_{eff}, which is big if ν_e are produced in very dense sites, as the center of the sun or a supernova core. The interplay between the vacuum and matter terms (MSW effect[5]) can make 'small' mixings important; let us discuss how, keeping in mind supernova neutrinos as a case study. At production, $|\nu_e\rangle$ is in good approximation the heaviest state, due to the new term in H_{eff}. If the mixing U_{e3} is large enough, ν_e remains always $|\nu_3\rangle$ ('adiabatic' conversion). Thence, neutrinos exiting the star have a very little overlap with $|\nu_e\rangle$, and $P_{e\to e} = |U_{e3}^2|$. Note that, even if U_{e3} is too small and fails to affect $P_{e\to e}$, U_{e2} is probably large enough to succeed, and $P_{e\to e} = |U_{e2}^2|$ can be already a large effect. (For antineutrinos, the effect goes in the opposite sense $|\nu_e\rangle \to |\nu_1\rangle$ and thence $P_{e\to e} = |U_{e1}^2|$.) Let us conclude with some remarks and qualifications: 1) most of the description above applies to supernova ν_e, since in the sun the electronic density is insufficient to overcome

---

[b]Unless stated otherwise, we will always consider a 'normal' mass spectrum, when $m_1 < m_2 \ll m_3$. However it should be recalled that existing data can be interpreted also assuming that $m_3 \ll m_1 < m_2$, a spectrum of mass called 'inverted' for obvious reasons.

the level separation of the big (atmospheric) Δm^2 (more discussion of supernova neutrinos below). 2) The MSW effect may be incomplete, and this is what is called 'non-adiabatic' case. In this case, the probability $P_{e \to e}$ is energy dependent. 3) The MSW effect can act also around the detection site (or whenever matter is crossed); e.g., solar neutrinos fluxes might be different in day and night (even if absorption in the Earth is negligible; we are discussing a coherent interaction of neutrinos with many electrons).

2. Investigating Massive Neutrinos With Oscillations

2.1. The Big Δm^2

As well known, and recalled by M. Shiozawa at this conference, there is a very strong evidence for oscillations of atmospheric neutrinos obtained by the Super-Kamiokande (SK) experiment. It should be recalled that MACRO results corroborate these claims, and a rather recent L/E analysis (in progress) adds further confidence in the interpretation. There are several interesting facets and some of them were discussed at the present conference: 1) the quantifications of the errors on atmospheric fluxes; 2) the puzzling results of Baksan; 3) the L/E distribution in Soudan 2 and the normalization of their electron neutrino flux. However, if one just browses the SK results, or if one compares 2 numbers quoted by Shiozawa $\chi^2_{\text{no osc.}} = 393.4/172$ (=something is wrong with standard picture at 9.5 σ) and $\chi^2_{\text{osc.}} = 157.5/170$ (=oscillations improve by 15 σ the agreement) one comes to the opinion that

$\boxed{\text{Atmospheric neutrino oscillate}}$

(or they behave very much alike) and that Δm^2_{32} and θ_{23} have been measured.[c] This formal statement corresponds to the fact that all data can be interpreted by 2 flavor oscillation between muon and tau neutrinos.

Main manifestation of oscillations in atmospheric ν's is muon *disappearance*; the production of τ lepton via charged current interactions is inhibited by the kinematical threshold.[d] However, data are so good that permit to ask a provocative question: whether we see ν_τ already. Indeed, SK obtained a *"NC enriched sample of events from multi-ring data"* (+NC for short). According to SK Montecarlo, 46 % of the +NC events are due to CC electron neutrino interactions, 25 % to CC muon neutrinos, and the remaining 29 % to NC. Pure disappearance of muon flux from below $f_\mu = F^{exp.}_\mu / F^{th.}_\mu \sim 1/2$ would imply a decrease of +NC events from below:

$$0.46 + 0.25 \cdot f_\mu + 0.29 \cdot \frac{1+2.1 f_\mu}{3.1}$$

---

[c]Indeed, the time seems mature to use atmospheric neutrino data to learn on atmospheric neutrinos themselves. This is relevant e.g. for the search of cosmic sources, of prompt neutrinos (from charm, or exotica), proton decay, ...

[d]τ appearance in atm. neutrino is limited by statistics: $\Gamma \approx 1 \, \tau$ *per* kton y. Thus, it seems difficult that SK will obtain a convincing evidence in a short time. τ appearance on long-baseline neutrino beam will be the target of OPERA and ICARUS. We do hope that Δm^2_{31} will not decrease too much in future analyses; present best values are around 2.5×10^{-3} eV^2.

This can be compared with the experimental up-down ratio, that is affected only weakly by uncertainties in the relevant cross sections:

$$\left.\frac{\text{Up}}{\text{Down}}\right|_{+\text{NC}} = \begin{cases} 1.28 & \text{disappearance with } f_\mu = 1/2 \\ 1.04 \pm 0.07 & \text{observed} \pm \text{ stat. error} \end{cases}$$

Number of σs is formal, but it is evident that pure disappearance is not the best hypothesis. Although this argument for ν_τ looks a bit indirect, and we may want to improve, one should note that this is just the same argument used to claim for appearance in solar neutrinos, based on SNO and SK data.[6]

The K2K experiment supports the evidence for disappearance, since they get 44 events when they expect 64. Final K2K data, and future investigations at MINOS and CNGS (with OPERA and ICARUS experiments) will most probably convince the whole community. As a general remark, and also in view of long-baseline experiments, we believe that it will be important to improve on neutrino interactions in the 1-100 GeV range. In this connection, it is interesting that the rate of production of π^0 via neutral current interactions has been already checked in the close Čerenkov detector of K2K experiment, and it agrees with the estimated number within 10 % (such a big error is perhaps due to the fact that the fiducial volume in that detector is just the tiny region of interest for long-baseline purposes; if this is the case, it could be interesting to think to dedicated analyses). Incidentally, these π^0 are the most important source of the +NC events discussed above.

The last question we want to address is: How to measure θ_{13} and $45° - \theta_{23}$?[e] We could get θ_{13} by a disappearance experiment: $P_{e \to e} = 1 - \sin^2 2\theta_{13} \cdot \sin^2 \varphi_{\text{atm}} \neq 1$ or the formula inclusive of MSW effects (when needed). Or, we could prove that $\theta_{13} \neq 0$ by appearance: $P_{\mu \to e} = \sin^2 \theta_{23} \cdot (1 - P_{e \to e}) \neq 0$; the ν_μ share is regulated by θ_{23} (seeing the remaining oscillated neutrinos by $P_{\tau \to e} = \cos^2 \theta_{23} (1 - P_{e \to e})$ is not a practical possibility).[f] With both results, we could perhaps tell $\theta_{23} > 45°$ from $\theta_{23} < 45°$. We cannot resist to propose a slogan: "don't judge only by appearances, use disappearances too". Let us conclude with some observations: (a) A bound $\sin^2 2\theta_{23} > 0.9$ is equivalent to the relatively weak bound $1/3 < \sin^2 \theta_{23} < 2/3$, simply due to trigonometry. Future search for ν_μ disappearance could eventually conclude that $\sin^2(2\theta_{23}) \neq 1$, but could not tell θ_{23} from $90° - \theta_{23}$. (b) The search of sign of $45° - \theta_{23}$ looks to be intertwined with (and more difficult than) the search of θ_{13}. (c) Note that also Δm^2_{21} may contribute to electron disappearance in long-baseline experiments, but in a way that is difficult to disentangle from possible effects of θ_{13}.[8]

2.2. The Small Δm^2

The evidence for physics beyond the standard model is above 7 σs. Note that Gallex/GNO alone gives more than 5 σs, using the predictions of the standard solar

---

[e]Note that we formulate the question on θ_{23} by accounting for the fact that data are compatible and perhaps suggest a maximal mixing, $\theta_{23} = 45°$.
[f]A recent work[7] estimates the sensitivity of MINOS to U^2_{e3} as 0.3 %, assuming $\sin^2 \theta_{23} = 1/2$.

model of Bahcall and collaborators. This anomaly survives important modifications of the standard solar model (however, there is no indication that this model has serious faults). The next generation of experiments (KamLAND and Borexino) together with existing experiments (and most remarkably SNO) will be able to identify the true solar-neutrino solution determining the solar-neutrino oscillation parameters. This is quite evident from a number of simulated fits, shown as fig.1 of a work of present authors[9]: the correct solutions are always identified[g] (though in certain regions of parameters, further study would be deserved to obtain an accurate measurements).

It should be noted that already present data suggest strongly that the solar mixing angle (to be identified with θ_{12} in good approximation) is large. Furthermore, a second Δm^2 (different from the atmospheric one, to be identified with Δm_{21}^2) has been uncovered. In most of the regions suitable to interpretation of solar neutrino data MSW effect is operative. Indeed, one could be tempted to conclude that the occurrence of MSW effect is a solid fact; however, one realizes at a closer look that there is a small region at $\Delta m^2 =$ few 10^{-10} eV^2 that has a very good χ^2 and undoubtly belongs to vacuum oscillation region (VO). How this region comes to exist? Going in the upper part of VO region, many oscillations are met and SK spectrum is increasingly flat. However, there is a competing pull toward lower values, where the beryllium neutrino flux is suppressed due $1.267\Delta m^2 L_\odot/E_{Be} \approx \pi/2$: this permits to accommodate the Homestake results better. The best compromise is obtained at $\Delta m^2 \approx 4.8 \cdot 10^{-10}$ eV^2; there, 4 oscillations are present in the high energy part of the spectrum (SK and SNO). In principle, SNO may check this specific solution since the electron energy correlates better with the neutrino energy; however, the energy resolution they quote does not permit to emphasize well wiggles in the spectrum with present data. In conclusion, there is no proof of MSW effect yet (note also that there are certain 'energy independent' solutions at big Δm^2).

3. Supernova Neutrinos

The only unknown mixing angle is θ_{13}; we only know from CHOOZ and SK that it cannot be much larger than 15°. However, θ_{13} can greatly affect the ν_e emitted from a supernova of type II due to MSW effect. Their flux F_e in the detector would have the distribution expected for the flux F_μ^0 of muon (or tau) neutrinos: $F_e \to P_{e\to e} F_e^0 + (1 - P_{e\to e})F_\mu^0 \approx F_\mu^0$. Since the expected average energy of ν_e is around 10 MeV, and the one of ν_μ is at least twice, one may suspect that these effects are observable.[h] As already mentioned, the effects of U_{e3} oscillations can be partially simulated by those related to U_{e2}-oscillations with 'solar' frequency. However, the main message we want to stress is that the effects of oscillation cannot be neglected when interpreting a supernova neutrino signal.

---

[g]In[9], the role of future sub-MeV experiments for neutrino mass measurement is also assessed.
[h]For fairness, it should be recalled that there are considerable uncertainties in the modelization and predictions of a supernova.

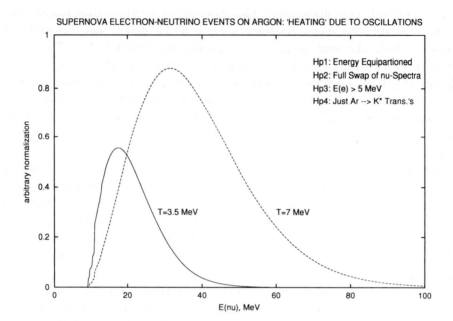

Fig. 1. Impact of oscillations on the electron neutrino signal at ICARUS.

To reveal the effect we are discussing, one should detect electron neutrinos. But, usually, supernova signal is dominated by $\bar{\nu}_e$ (*anti*-neutrino) on free proton reaction, which has a large cross section. There is a second class of reactions[i] that are useful for this purpose, for instance: $\nu_e + {}^{16}O \to e + {}^{16}F$ (with $Q = 15.4$ MeV) which can be exploited at water Čerenkov detectors (as SuperKamiokande or SNO) due to the angular distribution $P(\theta) = 1 - \cos(\theta)/3$;[j] or the reaction $\nu_e + {}^{12}C \to e + {}^{12}B$ (with $Q = 17.4$ MeV) which is double tagged due to the β^+ decay of Boron and this is useful in scintillators detectors (LVD, Borexino, KamLAND, BAKSAN); or the reaction (with large cross section): $\nu_e + D \to e + p + p$ (with $Q = -1.4$ MeV) that can be used at the inner part of SNO. The signal is given by a lone electron, in contrast with neutral current (or electron antineutrino) reactions on deuterium that are tagged by additional neutrons. However, there are some cases in which the electron neutrinos induce the dominant signal in a detector. This is for instance the case of the forthcoming ICARUS, which employs the reaction $\nu_e + {}^{40}Ar \to e + {}^{40}K$ (with $Q = 5.9$ MeV for the main nuclear transition) One has similar reaction on ^{37}Cl, though detectors as Homestake are not (yet?) instrumented for real time

---

[i]There are also other classes of reactions: those that measure neutral current events, that are just irrelevant to oscillations (though important in checking supernova neutrino fluxes), and those as the elastic scattering, that measure an admixture of charged and neutral current events. They offer the advantage of a well-known cross section and good directionality, and are of interest for oscillations. See the lecture of F. Cxi for a beautiful review.
[j]^{16}F rapidly decays by proton emission.

neutrino detection. In a sense, the detection of 'heated' ν_e requires to stay closer to the philosophy of solar neutrinos (=hunt for ν_e's), and to employ literally solar detectors. Note that a big Q value amplifies the difference between the case with and without oscillations. For the purpose of illustration, we show in Fig. 1 the effect on the number of events in ICARUS, plotted in function of the neutrino energy[10]. The energy spectra are assumed to be Fermi-Dirac, with the temperatures indicated in the figure. The low (high) temperature corresponds to (no) oscillation case.

4. Investigating Massive Neutrinos Without Oscillations

4.1. β Decay

At first sight, there is plenty of options for mass search using the β spectrum: (a) one may search effects at endpoint, when the momentum of neutrino is minimal and thence mass effects are emphasized; (b) one may search for kinks far from endpoint, which would inform us on 'heavy' neutrinos $m \sim$ fractions of keVs or more; (c) one may search for an overall shift of the spectrum due to the fact that production of neutrino mass takes away some energy. However, no signal seen yet; the best sensitivity to m_{ν_e} is 2-3 eV using the first method–'endpoint'[11]. This method may be pushed to $200 - 300$ meV (with an energy resolution of $\delta E_e \sim 1$ eV). It should be noted that oscillation scales $(\Delta m_{ij}^2)^{1/2}$ are much lower than this (except LSND, $300 - 1000$ meV). But there is an interesting (perhaps odd) case that could warrant discovery even for 3 neutrinos: if the mass splitting are small but at the same time all neutrino masses are big; indeed, we would have $m_{\nu_e} \approx m_i$ for all i. This case would be of particular interest for cosmology, since neutrino background radiation is left over since big-bang times.[12]

4.2. 0ν2β Decay and A New Bound for β Decay

The ee-element of the mass matrix $\mathbf{M}_{ee}$ triggers $0\nu2\beta$. Present best limit $350 \cdot h$ meV; may be improved to $20 \cdot h$ meV. h is a factor that quantifies the nuclear uncertainties, which presumably ranges in the interval $[0.6, 2.8]$. Let us assume the case suggested at the end of previous section. By playing with the Majorana phases, one realizes easily that there is a lower bound on $\mathbf{M}_{ee}$, namely:

$$|\mathbf{M}_{ee}| \geq \max(c_{13}^2 \cos 2\theta_{12} - s_{13}^2, 0) \times m_{\nu_e}$$

Thence, if we knew the angles well enough, we could convert a bound on mixing angles $|\mathbf{M}_{ee}|$ into a bound on m_{ν_e} (in other words, we can learn on β decay parameter from $0\nu2\beta$ and oscillations). Still, it is essential to know whether the allowed mixing angles permit a cancellation (so that the lower bound on $|\mathbf{M}_{ee}|$ reduces to zero) or not. In order to investigate these questions, we performed a χ^2 analysis where the existing oscillation data were merged with those from $0\nu2\beta$ decay[13]. The result is shown in Fig. 2. It should be noted that the bounds from this procedure are competitive with the existing ones, obtained by β decay experiments. A number of

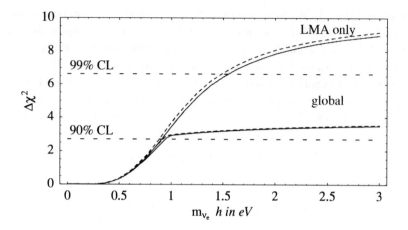

Fig. 2. The present experimental bound on the mass of degenerate neutrinos. The dotted lines illustrate the constraint obtained setting $\theta_{13} = 0$. $h \approx 1$ parameterizes the nuclear matrix element.

conclusive comments: (a) the argument given above stresses again the importance of improving on theoretical calculations of matrix elements (to reduce nuclear uncertainties); (b) if the $0\nu2\beta$ will progress as expected, it may give the most important bound on neutrino mass scale; (c) for this method to work, mixing angles must be known reliably. It is just essential to measure θ_{12} well; while θ_{13} is not crucial (dotted curves). (d) In the fortunate option that a signal will be seen in beta decay, and the mixing angles will be known well, the $0\nu2\beta$ decay will permit to study the Majorana phases. (For details, references, and other interesting possibilities probed by future $0\nu2\beta$ experiments see[14]).

5. Theoretical Approaches to Massive Neutrinos

We focus only on two rather specific models. The main goal is to illustrate how theoretical considerations may suggest (or guide) future search; however the specific choice among many model is just due to the personal involvement in their construction and/or discussion.

5.1. One Model Based on Flavor Symmetry

Maybe the structure of ν mass matrix is controlled by U(1) selection rules[15], e.g. ν_e has to pay one suppression factors—ν_μ and ν_τ instead not:

$$\frac{\mathbf{M}}{m_0} \stackrel{\mathcal{O}(1)}{=} \begin{pmatrix} \varepsilon^2 & \varepsilon & \varepsilon \\ \varepsilon & 1 & 1 \\ \varepsilon & 1 & 1 \end{pmatrix}$$

This can be tested with available information on ν's, assuming that the $\mathcal{O}(1)$ factors are random numbers[16]. Making connection with charged fermion properties, one can argue that ε is a power of $\sin\theta_C$. This yields $\theta_{13} \sim (|\mathbf{M}_{ee}|)^{1/2} \sim \sin\theta_C^n$ etc. It turns out that the case $n = 1$ is in reasonable agreement with data, indeed, there is a chance probability of ~ 10 % to get parameters that agree with phenomenology. LMA is often reproduced.

5.2. *One Model Based on SO(10)*

Program: We know charged fermion masses at M_Z; embed them in an SO(10) picture at M_{GUT}; then extrapolate back to M_Z, and learn on ν masses. The minimal renormalizable supersymmetric SO(10) model has 2 higgses:

$$W = -\mathbf{16}_i(\mathbf{10}\ Y_{ij}^{(10)} + \mathbf{126}\ Y_{ij}^{(126)})\mathbf{16}_j$$

Now: (1) the *singlet* in **126** gives huge mass to ν^c, (2) its *doublets* help correct **10**-plet predictions on fermion masses (**10**-plet dominance good only for 3rd generation), (3) its *triplet* can contribute to neutrino mass. The 2nd and 3rd generations Yukawa couplings at M_{GUT} are almost completely determined:

$$Y^{(10)} \approx \begin{pmatrix} 0.002 & 0 \\ 0 & 0.94 \end{pmatrix} \qquad \varepsilon\ Y^{(126)} \approx \begin{pmatrix} 0.014 & 0.027 \\ 0.027 & x \end{pmatrix}$$

Thus, if the triplet dominates the ν mass matrix, $\mathbf{M}_\nu \propto Y^{(126)}$: (*i*) θ_{23} can be large and (*ii*) weak mass hierarchies m_2^2/m_3^2 are favored (the larger θ_{23}, the weaker the hierarchy)[17]. Not a bad start in our view.

6. Conclusive remarks

Should we target discoveries or measurements? At present, this is unclear. Neutrinos have often showed surprising features in the past, and it is all too easy to argue that this might happen again. Here, we assumed that oscillations of atmospheric and solar neutrinos are established facts. In this view, we believe that next important and realistic experimental targets are the discrimination of solar neutrino oscillations parameters, and θ_{13}. The former seems a well defined program, if KamLAND and Borexino will work as we hope. Another implication of this view is that oscillations should be included for proper interpretation of neutrino signal from astrophysical sources (e.g. from supernova neutrinos).

Non-oscillations search may lead to surprises. However we stressed that within the hypothesis that neutrinos are massive particles, these results are related to oscillation searches. We discussed in particular how future $0\nu 2\beta$ search may play an important role on studies of the neutrino mass scale, with implications for β decay experiments.

Theory may help fill the gaps, but we feel that (at present) the best that can be done is to provide *patterns*, more than precise predictions. More pragmatic attitudes are to select very defined models as bimaximal mixing, to stay at phenomenological level, or to ignore theory alltogether. However, quark-lepton symmetry seems to be a good guide, and grand unification ideas remain tantalizingly attractive in connection with massive neutrinos.

Acknowledgments

F.V. would like to thank the Organizers for kind invitation at the workshop 'Matter, antimatter and dark matter', and F.Cei, A.Masiero, M.Shiozawa, T.Stanev, M.Vietri, and several other Participants (but especially G.Battistoni) for very pleasant discussions and good time had together at ETC* and in the beautiful town of Trento.

References

1. B. Pontecorvo, Sov. Phys. JETP **6** (1957) 429
2. S. M. Bilenkii, J. Hosek and S. T. Petcov, Phys. Lett. B **94** (1980) 495
3. M. Apollonio *et al.* [CHOOZ Collaboration], Phys. Lett. B **466** (1999) 415
4. For a recent discussion see A. Strumia, hep-ph/0201134
5. L. Wolfenstein, Phys. Rev. D **17** (1978) 2369; S. P. Mikheev and A. Y. Smirnov, Sov. J. Nucl. Phys. **42** (1985) 913
6. F. L. Villante, G. Fiorentini and E. Lisi, Phys. Rev. D **59** (1999) 013006
7. A. Para and M. Szleper, hep-ex/0110032.
8. R. Barbieri and A. Strumia, JHEP **0012** (2000) 016
9. A. Strumia and F. Vissani, JHEP **0111** (2001) 048
10. This result was obtained in collaboration with F. Cavanna and O. Palamara.
11. See e.g. the reports of MAINZ and TROITSK Collaboration in proceedings of *Neutrino 2000* Conference.
12. For phenomenological analyses, see H. Minakata and O. Yasuda, Phys. Rev. D **56** (1997) 1692; F. Vissani, hep-ph/9708483.
13. H. V. Klapdor-Kleingrothaus *et al.*, Eur. Phys. J. A **12** (2001) 147
14. F.Feruglio, A.Strumia and F.Vissani, 'What can we learn on Majorana neutrinos', to appear.
15. See e.g. T. Yanagida and J. Sato, Nucl. Phys. Proc. Suppl. **77** (1999) 293
16. F. Vissani, Phys. Lett. B **508** (2001) 79
17. B. Bajc, G. Senjanovic and F. Vissani, hep-ph/0110310 and references quoted therin

NEUTRINOS FROM SUPERNOVAE: EXPERIMENTAL STATUS

FABRIZIO CEI*

Dipartimento di Fisica dell' Università and INFN - Pisa
Via Livornese, 1291a, 56018 S. Piero a Grado (PI) - ITALY

I discuss the state of the art in the search for stellar collapse neutrinos and the perspectives of this field. The implications for neutrino physics of a high statistics supernova neutrino burst detection by the network of operating experiments are also reviewed.

Keywords: supernovæ; neutrinos; neutrino properties.

1. Introduction

After more than three decades of theoretical studies[1], there is now a rather general consensus about the basic mechanism of core collapse supernovæ[2]: the Iron core of a massive star overcomes its hydrodinamical stability limit and collapses, raising its density up to many times that of the nuclear matter; then, the core bounces elastically, producing a shock wave. The shock travels through the star, loses energy in dissociating nucleons and producing neutrinos and finally stalls, but is revived by ν_e, $\bar{\nu}_e$ absorption on nucleons ("neutrino heating"); so, it can reach the stellar surface and expel the external layers into the space. The star then cools down, mainly by neutrino emission, and degenerates into a neutron star or a black hole.

In this picture there are, however, still many question marks, as, for instance, the amount of energy supplied by the neutrino heating, the role of convection, star rotation and star magnetic field, the destiny of the residual and many others. The detection of the neutrinos emitted from a supernova (SN) could address (at least partially) such questions[3]; moreover, a galactic SN explosion would give particle physicists the opportunity to explore the neutrino properties on scales of distance up to $\sim 10^{17}$ Km, of time up to $\sim 10^5$ years and of density up to $\sim 10^{14}$ g cm^{-3}.

2. Supernova Neutrino Bursts

A SN explosion releases $\sim 2 - 4 \times 10^{53}$ erg of gravitational binding energy, the bulk of it in form of neutrinos. (The kinetic energy of the expelled matter and the energy emitted in light and gravitational waves are lower by at least two orders of magnitude.) The SN neutrino emission can be divided in two stages:

---

*E-mail: fabrizio.cei@pi.infn.it

Infall-Neutronization. A rapid ($\Delta t \sim 10$ ms) pure ν_e burst is emitted during the infall of the stellar matter above the core and few ms after the core bounce. The neutrinos are produced via electron capture on protons: $e^- + p \to \nu_e + n$; the average ν_e energy is ~ 10 MeV and the total released energy is $\lesssim 10^{52}$ erg.

Cooling. When the star cools down, neutrinos of all flavors are produced, via nucleonic bremsstrahlung and pair annihilation processes as: $e^+ + e^- \to \nu + \bar{\nu}$. Because of the huge density of the collapsed core, the neutrinos are trapped within it. Non-electron neutrinos and antineutrinos (from now on, collectively indicated with ν_x) interact with matter via neutral currents only, while ν_e and $\bar{\nu}_e$ interact via charged and neutral currents. Therefore, the ν_x scatter less frequently and decouple deeper inside the core than the other neutrinos; similarly, since the core is richer in neutrons than in protons, the $\bar{\nu}_e$ are less tightly coupled than the ν_e and emerge first. The core becomes hotter when the radius gets smaller; then, the ν_x have the highest mean energy (e.g.[4]): $\langle E_{\nu_e}\rangle \approx 10 - 13$ MeV; $\langle E_{\bar{\nu}_e}\rangle \approx 14 - 17$ MeV; $\langle E_{\nu_x}\rangle \approx 22 - 27$ MeV. The cooling stage lasts ~ 10 s and takes away $> 99\%$ of the gravitational energy of the star, roughly equally distributed between the neutrino flavors; the neutrino energy spectra are thermal (e.g. Fermi-Dirac) or quasi-thermal and can be characterized by a temperature T_ν and a chemical potential μ_ν.

3. Supernova Neutrino Reactions

The detection of SN neutrinos requires their conversion into a charged lepton or the emission of other particles (γ's or neutrons) which can be efficiently observed. The reactions employed in SN neutrino detection fall in four main groups.

Charged currents on nucleons (CCn(p)). The unique process of this cathegory which can be used is the inverse-beta reaction:

$$\bar{\nu}_e + p \to e^+ + n \tag{1}$$

since the ν_x energies are under the threshold for μ and τ production and there are no target materials containing free neutrons. The energy threshold of (1) is 1.8 MeV (well below the peak of $\bar{\nu}_e$ spectrum) and its cross section is proportional to the square of the positron energy: $\sigma_{\bar{\nu}_e p} = 8.5 \times 10^{-44} \, (E_{e^+}(\text{MeV}))^2$ cm^2 where $E_{e^+} \approx E_{\bar{\nu}_e} - 1.293$ MeV. A further signature of (1) is given by the thermalization and capture of the neutron by a proton, which forms a Deuterium nucleus and releases the binding energy as a 2.2 MeV photon (from now on, $\gamma_{2.2}$) with a typical delay of ~ 180 μs in respect with the positron. All SN1987A neutrinos observed by Kamiokande II, IMB3 and Baksan[5] are generally believed to be $\bar{\nu}_e$, detected via (1). This reaction has only a weak dependence on the incoming neutrino direction.

Elastic scattering on electrons (ES). This process is possible for neutrinos of all flavors; however, the cross section is larger for ν_e and $\bar{\nu}_e$ (more tightly coupled with matter): $\sigma_{\nu_e e} = 9.2 \times 10^{-45} E_{\nu_e}$ (MeV) cm$^2 \approx 3\sigma_{\bar{\nu}_e e} \approx 6\sigma_{\nu_x e}$. The angular distribution of the recoil electrons is peaked in the incoming neutrino direction, with an opening angle $\theta_{\nu,e} \sim (m_e/E_\nu)^{1/2}$. The Kamiokande, Super-Kamiokande[6] and SNO[7] experiments detected thousands of ES events induced by solar ν_e.

Charged currents on nuclei (CCN). Again, these reactions are possible only for ν_e and $\bar{\nu}_e$. Some CCN processes (e.g. on C) have a large cross section and a clean signature, since the product nuclei decay by $\beta^\pm$ or γ emission ($\tau \sim 20$ ms); however, the number of expected events is usually small because of the high energy threshold (~ 15 MeV for C or O), which cuts off most of the ν_e, $\bar{\nu}_e$ spectrum.

Important exceptions are the ν_e, $\bar{\nu}_e$-induced deuteron disintegration processes:

$$\nu_e + d \to e^- + p + p \tag{2}$$

$$\bar{\nu}_e + d \to e^+ + n + n \tag{3}$$

The energy thresholds for (2) and (3) are 1.44 MeV and 4.03 MeV, both well below the mean ν_e, $\bar{\nu}_e$ energy. The e^- in (2) and the e^+ in (3) are preferably emitted backward, but the forward-backward asymmetry is very weak. The process (2) was succesfully used by the SNO experiment to measure the solar neutrino flux[7].

An other interesting CCN reaction is:

$$\nu_e + {}^{40}\text{Ar} \to {}^{40}\text{K}^* + e^- \tag{4}$$

which has a 5.885 MeV energy threshold and is accompanied by the $^{40}\text{K}^*$ de-excitation to the ground state; the energy of the de-excitation photon is 5 MeV.

Neutral currents on nuclei (NC). As explained before, the ν_x detection is possible only via NC reactions. Two kinds of processes were suggested: the excitation and subsequent de-excitation of a nuclear level, with emission of photon(s), and the knocking-off of a neutron. All these processes are not sensitive to the neutrino energy and direction, but the neutrino arrival time can be measured.

Neutrinos can excite ^{12}C and ^{16}O nuclei. The de-excitation proceeds through the emission of a 15.1 MeV photon (from now on, $\gamma_{15.1}$) in the former case and of a $5 - 10$ MeV[8] e.m. cascade in the latter. Both these reactions have a high energy threshold and are therefore good selectors of the more energetic ν_x.

Neutrinos of all flavors can breakup the Deuterium nucles via the NC process:

$$\nu\,(\bar{\nu}) + d \to \nu\,(\bar{\nu}) + n + p \tag{5}$$

The energy threshold of (5) is 2.2 MeV, so that many Deuterium disintegrations are expected also from ν_e and $\bar{\nu}_e$. (The process (5) was indeed originally suggested to perform a flavor-independent measurement of the solar neutrino flux.)

Finally, some theoretical calculations (e.g.[9]) stressed the good ν_x selection capabilities of the neutron knocking-off processes from certain heavy nuclei (Ca, Na, Pb ...), whose cross sections steeply increase for $E_\nu \gtrsim 20$ MeV. The use of chemical compounds of high solubility in water, as Lead perchlorate $(\text{Pb}\,(\text{ClO}_4)_2)$, was also suggested to combine the large CCN and NC neutrino cross sections on high-Z nuclei with the experimental advantages offered by the water Čerenkov technology[10].

4. Present Supernova Neutrino Detectors

The galactic stellar collapses are very rare events (one over $20 - 40$ years[11]), the SN neutrino energies are not larger than tenths of MeV, their cross sections are small and, finally, a SN has (at least can have[a]) a bright optical flare and releases radiation in other forms than neutrinos; these points set the main guidelines in designing a good SN neutrino detector. First of all, it must have a mass $\gtrsim$ 1 kton of active material; secondly, it must be located underground to reduce the cosmic ray induced background, and possibly in a low radioactivity environment; then, it must have a very high duty cycle (in principle 100 %) and an operating life $\gtrsim$ 10 years and finally, it must be equipped with electronics and acquisition systems well suited for a real time neutrino detection, with (possibly) good angular and energy resolutions; its energy threshold should not exceed $\sim$ 10 MeV. The present SN neutrino detectors satisfy many (at least) of such requests. Note that most of them were build having in mind also other physical goals, as proton decay and magnetic monopole searches, solar and atmospheric neutrino detection, neutrino oscillation studies etc.

4.1. Water Čerenkov detectors

Water Čerenkov detectors use large volumes of highly purified water, equipped with an array of inward-looking PMTs. The energy and direction of relativistic charged particles can be inferred by the total amount of collected Čerenkov light and by the pattern of illuminated PMTs. The Čerenkov detectors have a continuous active volume, whose inner part is shielded from the external radioactivity background by the outer one; a fiducial volume can then be defined. The water Čerenkov detectors are mainly sensitive to (1), with few per cent contributions from other reactions.

The **Super-Kamiokande** experiment[12] (from now on, SK) is a 50 kton water Čerenkov detector located in the Kamioka mine (Japan) at a depth of $\approx$ 2700 m.w.e. and equipped with 11146 20″ PMTs; the photocathodic coverage is 40 % of the total surface and the fiducial volume for SN neutrinos is 32 kton. Energy and angular resolutions at 10 MeV are $\approx$ 16 % and 27°; the energy threshold is $\approx$ 6 MeV[b].

4.2. Heavy water Čerenkov detectors

SNO[13] is a large heavy water Čerenkov detector, located in the Creighton mine (Canada), at a depth of 6010 m.w.e.. This experiment is based on 1 kton of D_2O, observed by > 9000 PMTs and surrounded by a shield of 5 kton of H_2O (the inner 1.4 kton of water can also be used for SN neutrino detection). The energy and angular resolutions are close to that of SK; the energy threshold is $\sim$ 4 MeV.

SNO has a good sensitivity to neutrinos of all flavors thanks to the simultaneous presence of heavy and light water and to the versatility ensured by the Deuterium breakup reactions (2), (3) and (5). A good capture efficiency ($\approx$ 83 %[7]) for the

---

[a]The optical flare can be absent because the explosion "fizzles" or because the SN is in a sky region optically obscured by the halo luminosity or the cosmic dust.
[b]On 12 november 2001 a severe accident destroyed about half of the SK PMTs. The SK collaboration plans to equip the whole volume of the detector with the survived PMTs; this should degrade the resolutions by about a factor $\sqrt{2}$ and increase the energy threshold up to $\sim$ 8 MeV. The number of events expected from a galactic SN should be reduced by only few per cent.

neutron emitted in (5) will be obtained by adding Chlorine (in form of $NaCl$) to the heavy water; the Chlorine de-excitation produces a ≈ 8.6 MeV e.m. cascade.

4.3. Scintillation detectors

Scintillation detectors use large masses of high transparency mineral oils, segmented in hundreds of counters or enclosed in a container and looked by arrays of PMTs. Good timing (~ 1 ns) and energy ($\sim 10\%$ at 10 MeV) resolutions are the main quality factors of these experiments. The scintillation detectors are primarily sensitive to $\bar{\nu}_e$ via (1), with $5-10\%$ contributions from other processes (mainly NC on C). Thanks to their large light yield, these detectors are sensitive to $\gamma_{2.2}$.

The **LVD**[14] experiment (presently the largest operating scintillation detector) is located in the Hall "A" of the Laboratory Nazionali del Gran Sasso (LNGS), (Italy), at an average depth of 3100 m.w.e.. After many years of running with ~ 600 tonn of scintillator, LVD reached in 2001 its final mass of ≈ 1 kton, organized in 840 counters (1.5 m^3 each), interleaved with Streamer Tubes for cosmic muons tracking.

The **MACRO**[15] experiment, ended in December 2000 after 12 years of operation, with 570 tonn of active mass had a sensitivity close to that of LVD. MACRO was equipped with 476 long counters (≈ 12 m each) and Streamer Tubes.

The **Baksan** observatory[16], equipped with 200 tonn (fiducial volume) of scintillator, has been searching for galactic collapses since more than 20 years. In 1987 this detector recorded a 5 event burst, generally attributed to neutrinos from SN1987A[5].

Two scintillation detectors of the "egg-container" type should go on-line soon, **Borexino**[17] in the LNGS (Hall "C") and **Kamland**[18] in the Kamioka mine. Borexino has a multi-shielding structure, with a fiducial volume of 300 tonn of scintillator; Kamland has an active volume of ≈ 1 kton, surrounded by a veto of mineral oil and scintillator. Both detectors are well suited to detect high energy photons, as $\gamma_{15.1}$.

Table 1 shows the expected number of events in SK (32 kton of fiducial volume), SNO and LVD for a collapse at the Galactic Center (GC, distance 8.5 Kpc) based on the model[19]. ν indicates the sum of the contributions due to ν_e, $\bar{\nu}_e$ and ν_x.[c]

4.4. Other detectors

ICARUS[20] is a modular Liquid Argon projection chamber; the first module (active mass 600 tonn) will be installed soon in the LNGS (Hall "C"). The Liquid Argon is mainly sensitive to ν_e via (4); ~ 40 ν_e events are expected in one ICARUS module from a SN at the GC, with reasonable chances to detect the infall-neutronization burst; ~ 10 events are also expected by ES reactions.

The **High Energy Neutrino Telescopes** (**AMANDA** and **Baikal** (on-line), **NESTOR**, **ANTARES** and **NEMO** (under development)) are arrays of hundreds

---

[c]Since the cross sections for ^{16}O excitations were evaluated only recently[8] and since the average ν_x energy in this model is ≈ 16 MeV, the model[19] predicts no NC reactions on ^{16}O. Other models (e.g.[37]), based on harder ν_x spectra, predict ≈ 700 of these events in SK and ≈ 60 in SNO.

Table 1. Expected number of events in SK, SNO and LVD from a SN at the Galactic Center.

Reaction	SK Events	SK Fraction (%)	SNO Events	SNO Fraction (%)	LVD Events	LVD Fraction (%)
$\bar{\nu}_e + p$ (CCp)	7349	95.9	446	39.6	296	93.1
$\nu + e$ (ES)	199	2.6	46	4.1	9	2.8
$\nu_e + O$ (CCN)	50	0.65	4 − 5	0.35 − 0.44	-	-
$\bar{\nu}_e + O$ (CCN)	63	0.83	5 − 6	0.44 − 0.53	-	-
Total on O	113	1.5	9 − 11	0.8 − 1.0	-	-
$\nu_e + d$ (CCN)	-	-	113	10.0	-	-
$\bar{\nu}_e + d$ (CCN)	-	-	201	17.8	-	-
$\nu + d$ (NC)	-	-	311	27.6	-	-
Total on d	-	-	625	55.4	-	-
$\nu_e + C$ (CCN)	-	-	-	-	≈ 1	0.3
$\bar{\nu}_e + C$ (CCN)	-	-	-	-	≈ 1	0.3
$\nu + C$ (NC)	-	-	-	-	11	3.5
Total on C	-	-	-	-	13	4.1
Total	7661	100	1127	100	318	100

of Optical Modules (OM), deployed in long strings in deep sea or antarctic ice. Even if the energy threshold of these detectors is in the GeV range, they have some sensitivity to a galactic SN because of a collective effect, the excess of single counting rates produced in many OMs by thousands of low energy positrons. The **AMANDA** experiment is equipped with a SN trigger based on this concept[21].

The **radiochemical detectors**, as **GNO**, **SAGE** and **Homestake**, are sensitive to ν_e via CCN reactions. The active material is periodically ($T \sim 1$ month) extracted to look for isotopes produced by solar ν_e interactions. Since any information on neutrino time, energy and direction is lost, these experiments are of limited value as SN detectors; however, in case of a nearby SN a prompt extraction could allow to check whether a significant increase of the ν_e counting rate was observed.

4.5. *Ideas for future projects*

The success of experiments like SK or AMANDA and the demand of very large and refined neutrino detectors (needed for long baseline and neutrino factory projects) stimulated some ideas to extend the present detector technologies to much bigger scales. Many of these projects have good potentialities also as SN neutrino detectors.

UNO[22] is a project of a 650 kton (fiducial volume 445 kton) water Čerenkov detector. A galactic stellar collapse should produce $\sim 10^5$ e^+ in UNO, but also a SN in Andromeda (the galaxy closest to us, at a distance of 700 Kpc) should be observable in this detector, with a ~ 20 event signal. Since the SN rate in the Local Group is expected to be several times larger than that in our galaxy, an experiment capable to look beyond the Milky Way has a much higher chance of success.

SNBO/OMNIS[23] and **LAND**[24] are projects of large mass (~ 10 kton) detectors, made of high-Z materials (NaCl, Fe and Pb), whose main goal is the observation of NC events from a SN neutrino burst. The knocked-off neutrons should be observed by ^{6}Li or ^{10}B detectors or by Gd-doped liquid scintillators, interspaced within the target materials. In case of a galactic SN these detectors should observe $\sim 10^3$ events, mainly induced by ν_x. The OMNIS collaboration is considering also a different design, based on 2 kton of $(\text{Pb}(\text{ClO}_4)_2)$, with high sensitivity to ν_e too.

LANNDD[25] is a proposal of a 70 kton magnetized Liquid Argon tracking detector. LANNDD should observe ~ 3000 events (4) from a SN at the GC.

IceCube[26] is a project of a 1 Km^3 volume neutrino telescope, to be located at the South Pole, designed for the detection of ultra high energy neutrinos ($\gtrsim 10^{20}$ eV). This experiment should be equipped with a SN trigger of the AMANDA type, with improved sensitivity due to the larger (4800) number of OMs.

5. What can we learn ?

A complete review of all that might be learned from a SN explosion would require the space of a very long report; then I shall limit myself to few points.

5.1. *Supernova physics*

Supernova and neutrino parameters. Some parameters of the SN source and of the neutrino spectra could be extracted from the large neutrino signal from a galactic SN. The $\bar{\nu}_e$ temperature $T_{\bar{\nu}_e}$ and chemical potential $\mu_{\bar{\nu}_e}$ should be measurable with a 1 % accuracy fitting the spectrum of (1) events recorded by SK. A lower accuracy ($\sim 10\,\%$) measurement should be possible also for ν_e, via CCN reactions. The total number of ν_e and $\bar{\nu}_e$ events will provide two independent measurements of the source strength at the detector E_B/D^2 (E_B is the emitted energy and D is the SN distance), which could be compared to check the energy equipartition hypothesis. Since the NC reactions do not preserve the neutrino energy information, one can estimate T_{ν_x} only assuming the energy equipartition and comparing the observed number of NC events with that computed using the measured E_B/D^2. The rough estimation of T_{ν_x} might allow to test the temperature hierarchy: $T_{\nu_x} > T_{\bar{\nu}_e} > T_{\nu_e}$.

Fast supernova observation: the SNEWS. The astrophysical models predict, as confirmed by the SN1987A observation, that the neutrino signal preceeds the SN optical flare by some hours. The first light from a SN is very interesting for astronomers, since it carries information on the progenitor and its environment; so, many experiments developed systems for a prompt recognition of SN neutrino bursts. A detailed analysis is performed for every burst candidate to recognize whether it matches the expected characteristics for a true neutrino signal.

The monitoring systems of the various experiments (SK, LVD, SNO, MACRO (ended), AMANDA ...) are integrated in a coordinate network of SN detectors, the **SNEWS** (**S**uper**N**ova **E**arly **W**arning **S**ystem, Fig. 1), whose goal is to provide a fast alert to the astronomical observatories. SNEWS is based on a blind computer, which gets separate alarms from any participating experiment and looks for time coincidences, in a 10 s window, between such alerts. This procedure eliminates the human interventions and ensures a very high level of confidence, since an accidental coincidence of fake SN bursts between experiments at distances of thousands of Km is extremely unlike[27]. The estimated response time of SNEWS is $\gtrsim$ 10 minutes.

Identification of supernova direction. The neutrino detectors should provide to astronomical observatories not only the information that a SN exploded, but also some indications on where to look at it. The pointing back to the SN is possible using two different techniques: the angular distributions of some neutrino reactions and the triangulation between detectors, which takes advantage from SNEWS.

Angular distributions. The most promising reaction is the ES. With an expected signal of some hundreds of ES events, SK should be able to reconstruct the direction of the neutrinos from a SN at the GC with an accuracy $\delta\theta \lesssim 5°$ or even better, depending on the capability of separating the ES electrons from the dominant (1) signal. A lower accuracy localization can be obtained in SNO: $\delta\theta \approx 20°$; these results are largely independent from the details of the SN model[28]. Some direction information can be extracted also by (1) or by CCN reactions on ^{16}O or d.

The scintillation detectors, being sensitive to $\gamma_{2.2}$, can provide information on the SN direction looking at the few cm relative displacement between the reconstructed positions of the positron (1) (which is detected nearly where it is created) and of the neutron (which is boosted forward). The CHOOZ experiment demonstrated the feasibility of this technique using reactor $\bar{\nu}_e$[29]; for a galactic SN, the expected pointing accuracy of a scintillation detector with a SK-like mass is 9°.

Triangulation. The triangulation technique uses the difference in the arrival times of the neutrino signal on different detectors to determine the SN direction. This method does not look very promising, since the rather low statistics expected in many SN detectors cause large uncertainties in timing the beginning of the neutrino burst. For instance, the accuracy obtainable for a collapse at the GC using a triangulation between SK and SNO is only $\delta(\cos\theta) \gtrsim 0.5$[30]; however, an important improvement is expected from the application of this technique to (possible) future detectors, with higher statistics and sensitivity[21].

5.2. *Non-standard-model neutrino physics with supernovæ*

A galactic SN would provide important insights on non-standard-model properties of neutrinos, as neutrino masses and oscillations.

ν_e **mass.** The present $\bar{\nu}_e$ mass limit, obtained by Tritium β-decay experiments, is ≈ 3 eV[31]. A massless neutrino and a massive neutrino of energy E, emitted at the same time from a SN at a distance D, should arrive on the earth separated by:

$$\Delta t(E,m)\,(s) = 0.515\,(m\,(\text{eV})/E\,(\text{MeV}))^2\,D\,(10\,\text{Kpc}) \qquad (6)$$

The cooling phase duration (≈ 10 s) tends to mask this delay and determines the minimum mass whose effects can be explored. In case of SN1987A, model dependent and independent $m_{\bar{\nu}_e}$ limits were obtained in the range 11 – 23 eV[32]. However, taking advantage from the large signal expected in SK, this difficuly can be avoided using the events (~ 300) observed in the first 50 – 100 ms after the explosion[33]. A sensitivity of ≈ 3 eV should be reached, at the same level of the terrestrial bound.

ν_μ, ν_τ **masses.** The terrestrial limits of ν_μ and ν_τ masses are 170 KeV[34] and 18 MeV[35], hard to improve with usual techniques. A galactic SN should produce hundreds of ν_x events in the present detectors; therefore, several ideas were proposed (e.g.[9,23,36]) to set stringent ν_x mass limits (down to 50–150 eV) using SN neutrinos.

As an example of such ideas[37], assume that ν_e (and then $\bar{\nu}_e$) is massless while ν_τ and ν_μ (at least one) are massive, and consider two samples of events, the "Reference" $R(t)$, formed by massless neutrinos only, and the "Signal" $S(t)$, formed mainly by massive neutrinos, with some contamination from massless neutrinos. $R(t)$ could be extracted by SK (or LVD) $\bar{\nu}_e$ data or by the H_2O portion of the SNO signal and $S(t)$ by NC reactions on O in SK or on d in SNO etc; the contamination ($\sim 20\%$) comes from the $\bar{\nu}_e$ signal between 5 and 10 MeV in SK and from ν_e, $\bar{\nu}_e$ NC reactions on d in SNO. A finite ν_x mass is signalled by a statistically significant difference $\langle t_S \rangle - \langle t_R \rangle > 0$ between the average arrival times of the two samples. Fig. 2 shows the results of 10^4 simulations (each simulation is a SN at 10 Kpc) in the case of SNO. The upper part of the figure shows the distribution of $\langle t_S \rangle - \langle t_R \rangle$

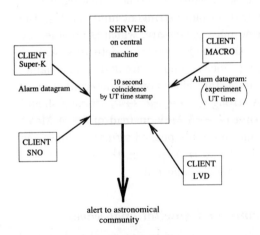
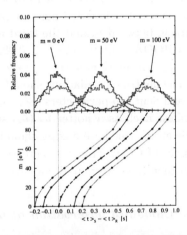

Fig. 1. The SNEWS setup.

Fig. 2. Mass sensitivity analysis for SNO. Upper part: distribution of the difference $\langle t_S \rangle - \langle t_R \rangle$ for $m = 0$, 50, 100 eV. Lower part: the range of masses corresponding to a given $\langle t_S \rangle - \langle t_R \rangle$. The meaning of the continuous, dashed and dotted lines is explained in the text.

for $m = 0$, 50, 100 eV; the distributions are narrower when $R(t)$ is taken from the larger statistics SK sample and broader when $R(t)$ is extracted from the SNO data. The lower part shows the central ν_x mass value (dashed line) corresponding to a given $\langle t_S \rangle - \langle t_R \rangle$; the continuous and dotted lines define the 10% and 90% C.L. limits obtained using SK data and SNO data for $R(t)$. Fig. 2 shows that the SNO sensitivity should reach ~ 30 eV, a six orders of magnitude improvement in m_{ν_τ}.

Similar analyses predict sensitivities down to 50 eV for SK, 55 eV for Kamland and 75 eV for Borexino[37,17]; all these results are obtained assuming that only ν_τ is massive. However, if, as suggested by the atmospheric neutrino anomaly[38], ν_μ and ν_τ are almost totally mixed, such sensitivities would improve by a factor $\sqrt{2}$. These results are quite robust, since the method uses the large $\bar{\nu}_e$ signal as an internal clock, without theoretical assumptions on the time pattern of the neutrino signal.

Even better sensitivities (down to 1.8 eV for ν_e and to 6 eV for ν_x) are expected in the particular case of a SN which rapidly degenerates into a black hole[39].

Neutrino oscillations. The SN neutrino spectra discussed in Sec. 2 are based on the hypothesis of massless neutrinos, which do not experience flavor mixing; however, important modifications are expected in case of neutrino oscillations.

The effects of neutrino oscillations on the SN neutrino spectra and their signatures in terrestrial detectors were discussed by many authors (e.g.[40]), taking into account also the earth-matter effects. Although some SN model dependence is unavoidable, the comparison between the observed spectra and that predicted using a set of oscillation parameters could allow to reject or accept the corresponding scheme. The general conclusions which can be drawn are that the analysis of the SN neutrino spectra could help to solve the solar neutrino problem, distinguish between normal or inverted mass hierarchy and set more stringent limits on U_{e3}, the coupling between ν_e and the third neutrino mass eigenstate, which is bounded by the CHOOZ result[41]: $|U_{e3}|^2 \lesssim 0.02$. Note that the SN neutrino detectors can not give any information on the $\nu_\mu \leftrightarrow \nu_\tau$ mixing, since ν_μ and ν_τ are undistinguishable.

The neutrino oscillations should have important effects on the determination of SN and neutrino parameters (Sec. 5.1). A strong $\nu_e \leftrightarrow \nu_x$ ($\bar{\nu}_e \leftrightarrow \bar{\nu}_x$) mixing should manifest itself as a hotter ν_e ($\bar{\nu}_e$) spectrum ($T \sim 8$ MeV instead of $3 - 4$ MeV) while an intermediate mixing should produce a doubly peaked structure. Since the number of CC reactions is increased by harder ν_e ($\bar{\nu}_e$) spectra, the normalization (and then the determination of E_B/D^2) will be also affected.

5.3. *Combined observation of neutrinos and gravitational waves*

Supernovæ are expected to emit not only neutrinos and light of various frequencies, but gravitational waves (GW) too[42], although the properties of the gravitational pulse are rather uncertain; the frequency range, for instance, extends from ~ 100 Hz to ~ 10 KHz. The GWs are even less coupled with matter than the neutrinos; so, they come from the deep interior of the star and do not experience a slow diffusion in the SN core. Then, a combined observation of neutrinos, GWs and (possibly) light from a SN would provide a very comprehensive picture of the collapse mechanism. Moreover, the gravitational pulse could be used to time the start of the collapse; this opens the possibility to set ν_e mass limits at the level of fractions of eV[43].

Many GW detectors of the resonant bar type (**EXPLORER, NAUTILUS, AURIGA, ALLEGRO, NIOBE**) are on-line and already sensitive to galactic SN, and many of the interferometric type (**VIRGO, LIGO, AIGO**) are under construction. An absolute time accuracy of a fraction of ms (needed for a good correlation) looks within the reach of such experiments. The interferometers are sensitive to frequencies from few Hz to tenths of KHz, while the bar detectors have a narrow frequency band ($\Delta f \sim 50 - 100$ Hz), centered at ~ 1 KHz. Efforts are under way to enlarge the bar detector bandwidths; however, the observation of an optical pulsar with 467.5 Hz emission frequency in SN1987A[44] stimulated the AURIGA people to try to tune the central frequency of their bar to this value[45].

6. Conclusions

The next galactic SN will be a huge source of physical and astrophysical information. The multi-flavor neutrino signal provided by the network of operating detectors will allow detailed studies of many aspects of the collapse mechanism; correlations with optical and GW observations will make the collapse picture more complete. Astronomers will be helped by the early alert provided by SNEWS and by the fast localization obtained by the event angular distributions and (maybe) by the triangulation between various detectors. At the same time, the particle physicists will have a powerful source of neutrinos, with exciting possibilities to set stringent limits on ν_x masses and on neutrino mixing. The fact that many of the results discussed here are weakly model-dependent makes this opportunity even more appealing.

However, a basilar *"caveat"* is needed: since there are still many obscure points and uncertainties in the SN neutrino signals and the experiments have, in any case, a limited statistics, the risk of rushed conclusions is just around the corner.

Acknowledgements

First of all, I want to thank all the members of the MACRO Collaboration and particularly my senior and younger Pisa colleagues (A. Baldini, C. Bemporad, M. Grassi, D. Nicolò, R. Pazzi and G. Signorelli) for their precious suggestions and criticism. Then (last, but not least), I want to thank the Organizing Committee of this workshop and expecially G. Battistoni for inviting me to give this review talk.

References

1. S. A. Colgate and R. H. White, *Ap. J.* **143**, 626, (1966); H. Bethe, *Rev. Mod. Phys.* **62**, 801, (1990); J. R. Wilson and R. Mayle *Phys. Rep.* **227**, 97, (1993); T. A. Thompson and A. Burrows, *Nucl. Phys.* **A688**, 377, (2001); A. Mezzacappa *et al.*, *Phys. Rev. Lett.* **86**, 1935, (2001); G. G. Raffelt, astro-ph/0105250
2. A. Mezzacappa *et al.*, astro-ph/0010580
3. M. Prakash *et al.*, astro-ph/0103095; T. Totani *et al.*, *Ap. J.* **496**, 216, (1998)
4. H. T. Janka in *"Frontier Objects in Astrophysics and Particle Physics"*, Vulcano (Italy), (1993)
5. IMB Coll., *Phys. Rev. Lett.* **58**, 1494, (1987); Kamiokande Coll., *Phys. Rev. Lett.* **58**, 1490, (1987); Baksan Coll., *JETP Lett.* **45**, 589, (1987)
6. Kamiokande Coll., *Phys. Rev. Lett.* **77**, 1683, (1996); E. Blaufuss (Super-Kamiokande Coll.), *"ICRC2001"* Conference, Hamburg (Germany), 2001
7. C. Waltham (SNO Coll.), *"ICRC2001"* Conference, Hamburg (Germany), 2001
8. M. Langanke *et al.*, *Phys. Rev. Lett.* **76**, 2629, (1996)

9. D. B. Cline et al., *Phys. Rev.* **D50**, 720, (1994)
10. S. R. Elliot, *Phys. Rev.* **C62**, 065802, (2000)
11. G. A. Tammann et al., *Ap. J. Suppl.* **92**, 487, (1994); S. Van den Bergh, *Comments Astrophysics* **17**, 125, (1993)
12. Super-Kamiokande Coll., *Phys. Rev. Lett.* **81**, 1158, (1998)
13. SNO Coll., *Nuclear Instr. and Meth.* **A449**, 172, (2000)
14. LVD Coll., *Nuovo Cimento* **105A**, 1793, (1992); A. Vigorito (LVD Coll.), "ICRC2001" Conference, Hamburg (Germany), 2001
15. MACRO Coll., *Astropart. Phys.* **1**, 11, (1992); MACRO Coll., *Astropart. Phys.* **8**, 123, (1998)
16. Baksan Coll., *Phys. Nucl. Part.* **29**, 254, (1998)
17. L. Cadonati et al., hep-ph/0012082
18. R. Svoboda (Kamland Coll.), "ICRC2001" Conference, Hamburg (Germany), 2001
19. A. Burrows et al., *Phys. Rev.* **D45**, 3362, (1992)
20. ICARUS Coll., hep-ex/0103008
21. F. Halzen et al., *Phys. Rev.* **D49**, 1758, (1994); T. Neunhöffer (AMANDA Coll.), "ICRC2001" Conference, Hamburg (Germany), 2001
22. C. K. Jung, hep-ex/0005046
23. D. B. Cline et al., *Astrophysical Letters and Communications* **27**, 403, (1990); P. F. Smith, *Astropart. Phys.* **8**, 27, (1997)
24. C. K. Hargrove et al., *Astropart. Phys.* **5**, 183, (1996)
25. D. B. Cline et al., "KEK-NUFACT 01 Neutrino Factory meeting", 2001
26. A. Goldschmidt (IceCube Coll.), "ICRC2001" Conference, Hamburg (Germany), 2001
27. K. Scholberg in "Amaldi Conference on Gravitational Waves", Caltech, 1999
28. J. F. Beacom and P. Vogel, *Phys. Rev.* **D60**, 033007, (1999)
29. CHOOZ Coll., *Phys. Rev.* **D61**, 012001, (2000)
30. J. F. Beacom, hep-ph/9909231
31. C. Weinheimer et al., *Phys. Lett.* **B460**, 219, (1999); V. M. Lobashev et al., *Phys. Lett.* **460**, 227, (1999)
32. J. N. Bahcall, "Neutrino Astrophysics", Cambridge University Press, (1989)
33. T. Totani, *Phys. Rev. Lett.* **80**, 2039, (1998)
34. K. A. Assamagan et al., *Phys. Rev.* **D53**, 6065, (1996)
35. ALEPH Coll., *Eur. Phys. J.* **C2**, 395, (1998)
36. A. Acker et al., *Phys. Lett.* **238**, 117, (1990); L. M. Krauss et al., *Nucl. Phys.* **B380**, 507, (1992); O. Ryazhskaya, *JETP Lett.* **56**, 417, (1992); G. Fiorentini and C. Acerbi, *Astropart. Phys.* **7**, 245, (1997)
37. J. F. Beacom and P. Vogel, *Phys. Rev.* **D58**, 093012, (1998)
38. Super-Kamiokande Coll., *Phys. Rev. Lett.* **81**, 1562, (1998); MACRO Coll., *Phys. Lett.* **B434**, 451, (1998); Soudan Coll., *Phys. Lett.* **B391**, 491, (1998)
39. J. F. Beacom et al., *Phys. Rev. Lett.* **85**, 3568, (2000)
40. Some examples of recent analyses are: G. M. Fuller et al., *Phys. Rev.* **D59**, 085005, (1999); A. S. Dighe and A. Y. Smirnov, *Phys. Rev.* **D62**, 033007, (2000); S. Choubey and K. Kar, *Phys. Lett.* **B479**, 402, (2000); C. Lunardini and A. Y. Smirnov, hep-ph/0106149; G. Fogli et al., hep-ph/0111199; K. Takahashi et al., hep-ph/0105204; F. Vissani, this workshop
41. CHOOZ Coll., *Phys. Rev. Lett.* **B466**, 415, (1999)
42. B. F. Schultz in "Encyclopedia of Astronomy and Astrophysics", London, 2000
43. M. Arnaud et al., hep-ph/0109027
44. J. Middleditch et al., astro-ph/0010044
45. C. Bemporad, private communication

Dark Matter and Gamma Rays
Chairperson: B. Bertucci

SEARCHES FOR DARK MATTER PARTICLES THROUGH COSMIC RAY MEASUREMENTS

PIERO ULLIO*

SISSA, via Beirut 4, 34014 Trieste, Italy

We consider the hypothesis that dark matter is made of weakly interacting massive particles (WIMPs) and describe how their pair annihilation in the galactic halo generates exotic cosmic ray fluxes. Features for generic WIMP models are reviewed, pointing out cases in which clear signatures arise. Implications from available and upcoming measurements are discussed.

Keywords: Dark matter; WIMP; cosmic rays.

1. Introduction

The fast progresses occurred in the latest few years will probably be remembered as those marking our entrance in the age of precision cosmology: observations spanning from galactic scales to the largest scales currently possible to probe, see, e.g.,[1], are converging on a picture with a flat Universe in which about 70% of the energy density is in some dark energy form, while 30% is provided by matter, most of which turns out to be dark, non-baryonic and cold (i.e. non-relativistic at the time when structure formation started).

The hypothesis that dark matter is made of weakly interacting massive particles (WIMPs) is particularly appealing, as it conjugates motivations from cosmology and from particle physics. WIMPs arise naturally in theories beyond the standard model of particle physics. Supersymmetry (SUSY), which seems a necessary ingredient to unify gravity with the other fundamental forces, provides the leading WIMP dark matter candidate: the lightest neutralino - a linear combination of the supersymmetric partners of the photon, the Z boson and neutral scalar Higgs particles, and hence a particle with zero electric and color charges - is, in large regions of the parameter space, the lightest supersymmetric particle, a stable particle in R-parity conserving SUSY models. If the particle physics model we are considering contains a stable massive particle (which we indicate here generically with χ) coupled to ordinary matter, we expect to find such particle in thermal equilibrium in the early Universe, and eventually make up relic populations nowadays. Two competing effects determine the relic abundance of χs: the χ pair annihilation rate into lighter

---

*E-mail: ullio@sissa.it

particles maintains χs in thermal equilibrium, until the Universe expansion takes over and makes the cosmological density of χs to freeze in. To first order[2], the relic abundance for a WIMP scales as:

$$\Omega_\chi h^2 \simeq \frac{3 \cdot 10^{-27} \text{cm}^{-3}\text{s}^{-1}}{\langle \sigma_A v \rangle}, \qquad (1)$$

where $\langle \sigma_A v \rangle$ is the thermally averaged total annihilation cross section. Now the argument closes up because, if we assume $\Omega_\chi h^2$ to be in the range currently favored for cold dark matter, about $0.1 - 0.2^3$, we find that the coupling of χ to ordinary matter has to be of weak interaction strength.

The scheme we have just sketched was understood over two decades ago, and it is still the most popular; it is usually referred to as the case of "thermal" relic WIMPs. Several other plausible scenarios have been studied more recently, see, e.g.,[4]: WIMP dark matter candidates which arise in these schemes are generically indicated as "non-thermal", as they rely on an alternative mechanism to provide the cold dark matter we see in the Universe today. Some degree of fine-tuning is usually required; at the same time, however, for these dark matter candidates phenomenological implications may be even sharper than for standard thermal WIMPs, including the possibility of very large pair annihilation cross sections, which, as we will see, are very interesting from the point of view of indirect detection of dark matter through cosmic ray searches.

2. WIMP annihilations an exotic cosmic-ray source

According to the current theory of structure formation, the ΛCDM model, cold dark matter drives the formation of all the structures we observe in the Universe today, see, e.g.,[5]: small dark matter halos merge, in a hierarchical fashion, into larger and larger dark structures; baryons, which are initially smoothly distributed, collapse in halo potential wells, cool and give rise to luminous components. If WIMPs are indeed the relics accounting for dark matter, then we expect them to populate all dark halos (and in particular the halo of our own Galaxy) with a distribution that follows the radial density profiles found with N-body simulations[6,7] (maybe with slight modifications due to the interplay with baryons, which are not included in the simulations yet) and with velocities of the order of halo circular velocities (about 200 km s^{-1} for a spiral of the size of the Milky Way). Although such densities and velocities are much smaller than in the early Universe, there is still a finite probability for WIMPs to annihilate in pairs: This effect is negligible from the point of view of the depletion of dark matter particles (except maybe in the center of very singular dark matter profiles or dark matter clumps) but maybe significant from the point of view of injecting in the hosting galaxy exotic primary cosmic rays[8,9].

WIMP annihilations generate quarks, leptons, gauge bosons and Higgs bosons, which in turn fragment and/or decay into stable species we might be able to identify in cosmic ray measurements. Details in annihilation rates and in branching ratios, and hence in the type and abundance of the generated species, depend on the specific particle physics model one considers; e.g., in the minimal supersymmetric extension to the Standard Model (MSSM), the relevant tree-level two-body final states from neutralino annihilations in the halo are: $c\bar{c}$, $b\bar{b}$, $t\bar{t}$, $\tau^+\tau^-$, W^+W^-, Z^0Z^0, W^+H^-, ZH_1^0, ZH_2^0, $H_1^0H_3^0$ and $H_2^0H_3^0$, with branching ratios rather sensitive to the free parameters in the model, but generally with heavy quarks or gauge bosons as dominant channels. Other combinations of Higgs and gauge bosons violate CP and are excluded; final states with light fermions are allowed but severely suppressed, as for any massive non-relativistic Majorana fermion (such as the neutralino in halos) the S-wave annihilation cross section into a fermion-antifermion pair is proportional to the fourth power of the mass of the fermion in the final state. Disregarding such details, the main features in the process of cosmic ray generation are common to any WIMP model. First, for thermal relic WIMPs, the strength of the total annihilation cross section at zero relative velocity $(v\sigma)_{v=0}$ (the appropriate limit for particles in galactic halos) is not free to vary over several orders of magnitude; in most cases (i.e. far from thresholds and resonances [10], and neglecting the case of so called coannihilations [10,11]) it is expected to be of the same order as for the thermally averaged annihilation cross section, fixed by the WIMP relic abundance, see Eq. (1). For non-thermal WIMPs the band of variation can be larger, but, as we already mention, most often it turns out that the annihilation cross section is enhanced rather than suppressed. At the same time, it is very hard to define a particle physics model in which the relative strength of the various cosmic ray yields is drastically different from what obtained in another sample case: most particles are not generated promptly, but arise in fragmentation processes or decay chains initiated by heavy tree level final states, common to all species. This should be kept in mind when testing a given model against a set of cosmic ray data: the comparison with data on other cosmic ray species is always needed and, if the model fits one set of data but severely violates a bound in another channel, it is very hard to define an alternative particle physics model which escapes such bound.

The source function for the species s can be written as:

$$Q_s^\chi(E,\vec{x}) = (v\sigma)_{v=0} \left(\frac{\rho_\chi(\vec{x})}{m_\chi}\right)^2 \sum_f \frac{dN_s^f}{dE} B^f \qquad (2)$$

where for a given annihilation channel f, B^f and dN_s^f/dE are, respectively, the branching ratio and the energy distribution for the yield s. The source function scales with the square of the WIMP number density locally in space, which is equal to the dark matter density divided by the WIMP mass: Any local enhancement in the dark matter density $\rho_\chi(\vec{x})$ (which might be due to the existence of halo profiles singular towards the galactic center or clumps in the halo, as indicated by N-body simulations) corresponds to a sharp increase in the cosmic ray yield, see, e.g,[12]. On the other hand, the scaling with the WIMP mass makes the detection of heavy WIMPs harder (a slight suppression for heavy masses is implicitly contained in the annihilation cross section as well).

The stable cosmic ray species generated in WIMP annihilations include gamma-rays, positrons, neutrinos, antiprotons and antideutrons, and, in the same amounts, their counterparts with opposite lepton or baryon numbers. The focus is generally on gamma-rays and antimatter, as in these cases the backgrounds from standard sources are well understood and/or scarce. Neutrino fluxes from WIMP annihilations in the galactic halo are negligible; an exception may be the Galactic center in case a huge density enhancement is present there[13] (but see[14]; recall also that the search for neutrino fluxes from the annihilation of WIMPs that are gravitationally trapped in the center of the Sun or the Earth is another very promising technique to detect WIMP dark matter[15]).

Gamma-rays propagate along geodesics, generally with very little absorption, and hence the flux observed at Earth is just the sum of contributions along a given line of sight. If the induced flux is indeed at the level of being measured, gamma-ray surveys are probably your only chance of mapping the fine structure of dark matter halos, with a much better resolution than what will be possible to achieve with strong lensing images (which have provided the first direct maps of dark matter distributions, see, e.g., [16]), maybe at the level of verifying whether the current picture from N-body simulations is correct. Among proposed dark matter gamma-ray sources, there are the Galactic center, the whole Milky Way halo (in case the dark matter distribution is clumpy), external galaxies[17] and cosmological sources[18]. The case for charged particles is slightly more involved as they diffuse in galactic magnetic fields; with more and more data available and with a wider range of species detected, the model for the propagation of charged cosmic rays in the Milky Way halo is being refined [19,20], and it is not any more a source of large uncertainties. The flux measured at earth is dominated by local sources, therefore with charged cosmic rays one may hope to probe the dark matter distribution in a nearby portion of the Milky Way (obviously, with no directionality), slightly larger with antiprotons rather than with positrons that lose energy more efficiently. Finally, a more indirect (and model dependent) tool to test the emission of electrons and positrons in WIMP annihilations in far away sources immersed in a magnetic field is to search for synchrotron radiation[21]; this method has been studied to search for dark matter in clusters of galaxies, in the Galactic center region, and, very recently, to identify dark matter clumps in the Milky Way halo[22].

The presence of large fluxes providing sizeable excesses is a necessary but not sufficient condition for dark matter detection. In most cases the estimate of the background is not entirely solid; to make a definite statement about the nature of a given excess, excluding more mundane sources, a clear signature for the dark matter induced component is needed. We review below the cases in which such signatures are present; in all other cases the comparison between fluxes and data will just provide upper limits for the exotic component, which in turn can be translated into joint constraints on WIMP models and dark matter distributions.

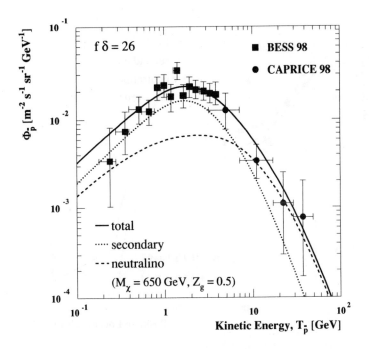

Fig. 1. Sample spectrum for the WIMP induced antiproton flux. The dark matter candidate considered here is a neutralino in the anomaly mediated supersymmetry breaking scheme[4], with mass equal to 650 GeV and gaugino fraction $Z_g = 0.5$. The dark matter halo is assumed to be clumpy with a moderate clumpyness factor, $f\delta = 26$, see[12] for details.

3. Spectral signatures of dark-matter induced cosmic-ray fluxes

The production of antiprotons in WIMP annihilations was the first case discussed in the literature in some detail. It is usually assumed that there is no standard primary astrophysical source of antimatter. Antiprotons can be generated in the interaction of primary cosmic rays (mainly protons) with the interstellar medium (mainly hydrogen atoms); looking at the kinematics in such process, one finds that the antiprotons tend to be produced with non zero momentum. There is no a priori reason to have such suppression at low kinetic energies for antiprotons from WIMP annihilations: Indeed, depending on WIMP mass and composition, the peak of the flux in this exotic component may be at energies much lower that for the standard secondary component. This signature[8,9] was exploited to claim dark matter detection when early measurements of the low energy antiprotons reported abnormally large fluxes[23]. Unfortunately, more recent and accurate measurements [24,25] have not confirmed the early data, and are now consistent with fluxes of secondary origin alone. Although an exotic component is not ruled out, the low energy signature

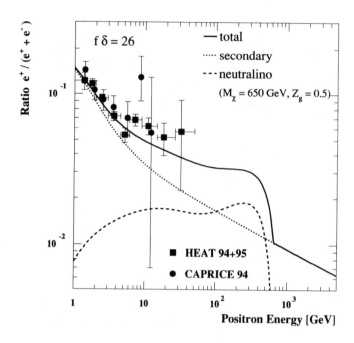

Fig. 2. Sample spectrum for the WIMP induced positron flux. The dark matter candidate considered here is the same as in fig.1.

is fading out. An alternative signature[26], a break in the antiproton spectrum at energies above few tens of GeV, may appear for heavy WIMP dark matter candidates in case the halo of the Milky Way is substantially clumpy. Such break due to a tail of antiprotons produced with very large momenta cannot be mimicked in conventional schemes.

The flux of antideuterons[27], formed from the merging of an antiproton and an antineutron, is expected to be about four orders of magnitude smaller than for the antiproton flux; nevertheless, the ratio signal to background at low energies may be much higher than for antiprotons. If future instruments achieve the required sensitivities, antideuteron measurements may set severe constraints or, eventually, allow for dark matter detection[27,28]. A few model dependent assumptions are involved in these conclusions, nevertheless this detection technique has several intriguing aspects and certainly deserves further investigations.

In case gauge bosons or leptons are the dominant channels in WIMP annihilations, the induced positron flux is generated in prompt decays, such as the decay of W^+ and τ^+ [29,30], and with a very distinctive spectral signature: The flux is peaked at an energy equal to about one half of the WIMP mass m_χ, while at higher energies it is sharply suppressed (and it is zero above m_χ). Upcoming experiments[31,32] will search for such a decrease in the range few tens to few hundreds of GeV. Note, however, that if the positron flux is instead generated in fragmentation processes initiated by WIMP annihilations into quarks, and with pions and kaons in the intermediate states, the peak in energy spectrum is shifted to lower energies and becomes rather featureless, definitely more problematic from the point of view of dark matter detection.

About 20-30% of the energy released in WIMP annihilations goes into gamma-rays. Most of them (about 90%) are generated in the decay of neutral pions produced in fragmentation processes[33]. The photon energy spectrum from $\pi^0 \to 2\gamma$ has a distinctive shape, the so called "π^0 bump": in the rest frame of the pion, the energy of each γ is equal to the mass of π^0 divided by 2; by making a Lorentz transformation, one can show that in the laboratory frame (i.e. the physical frame in our process) the γ-ray spectrum, if plotted on a logarithmic scale, is flat and symmetric around $M_{\pi^0}/2$. Spectra extending to higher energies correspond larger pion energies: the "bump" results then from the piling up of contributions from pions with different energies around the peak at $M_{\pi^0}/2$, see, e.g.,[34]. Unfortunately, the π^0 intermediate state is not unique to the dark matter induced flux; on the contrary it is common to many astrophysical processes, including the diffuse γ-ray production in the interaction of cosmic rays with the interstellar medium, the main background one has to fight against in disentangling dark matter sources. If one does know the characteristic spatial distribution of dark matter particles, one could use such piece of information as an angular signature to discriminate against the background; on the other hand, if the γ ray energy spectrum is measured with sufficient accuracy in the range around 100 GeV or so, a spectral signature may be singled out: the spectrum of the pions produced in WIMP annihilations is characterized by a single energy scale, the WIMP mass; it is peaked at $M_{\pi^0}/2$, dies out rather rapidly approaching M_χ and does not settle on any spectral index. Background sources cannot be characterized by a single energy scale, the flux arises as a superposition of terms weighted over some spectral index, the same spectral index the γ-ray flux settles on at energies much larger then $M_{\pi^0}/2$ (above a few GeVs). A dark matter induced component may then be singled out in case the measured flux can be safely decomposed in a contribution characterized by a spectral index plus a term with a fixed energy scale, with the total flux showing a spectral break at M_χ.

Finally, if two body final states containing a photon are non-negligible, a tiny fraction of the energy released in the annihilation may go into a gamma-ray flux with a much better signature: Although not allowed at tree level, essentially by definition of WIMP, such states are present at 1-loop level and are interesting because, as WIMPs in the galactic halos move with non-relativistic velocities, $v/c \sim 10^{-3}$, the outgoing photons are nearly monochromatic, with energy of the order of the neutralino mass[35,36,37]. There is no other known astrophysical source showing a similar feature: the detection of a line signal out of a spectrally smooth gamma-ray background would be a spectacular confirmation of the existence of dark matter in form of exotic massive particles.

4. Induced fluxes versus current data and future prospects

Current cosmic ray data do not show any evidence for an exotic component due to WIMP annihilations. In the energy ranges of interest to search for such an effect, the most accurate available measurements are those for antiprotons. At the moment, the antiproton channel is also the one in which the ratio signal to measured background is the highest, of order 1 for the most favourable cases of thermal WIMPs in the MSSM and considering a smooth halo scenario. Existing data can be used to rule out models for WIMP dark matter, otherwise allowed both by particle physics data and by other astrophysical measurements[38].

An indication of an excess with respect to the standard prediction of the secondary component has been found in positrons measurements[39,40]. This excess opens up a window for a WIMP induced component[41]; note however that the ratio signal to background is generally much smaller than for antiprotons, up to about 1/10 for thermal WIMPs in the MSSM and no dark matter clumps in the solar neighborhood. Typically, local enhancements in the dark matter density are needed for the exotic positron flux to be detectable and most probably, if a WIMP signal is indeed present in the positron measurements, an exotic component should be singled out in the antiproton flux as well, compare Fig. 1 with Fig. 2.

Excesses compatible with radiation from WIMP annihilations have been claimed in gamma-ray measurements as well. Data from the EGRET telescope (energies between few tens of MeV to 20 GeV) seem to show these features in the diffuse gamma-ray flux towards the Galactic center[42] and at high latitudes[43]: although dark matter sources might account for them, alternative explanations have been risen as well[44] and the data may just show that we need a better understanding of the model for the diffuse gamma-ray emission in the Galaxy (but it is rather puzzling that, in all attempts so far, ad-hoc assumptions are always needed to reconcile gamma-ray and charged cosmic ray data, such as, e.g., the hypothesis that the electron flux measured at earth is dominated by some nearby source rather than being representative for the Galactic primary electron flux). Note also that for a Galactic dark matter induced gamma-ray flux to be at the level of the excesses claimed from EGRET, again sensible enhancements in the dark matter densities are needed; in particular the hope of being able to identify dark matter by mapping in gamma-rays the Galactic center depends critically on whether the dark matter profile is singular towards it or not[45].

A reassuring piece of news is that the picture will be soon much clearer, with a much wider set of data that is going to be available. Data from upcoming space experiments[31,32] will measure antiprotons and positrons (and maybe antideutrons) with a much better statistics and energy coverage, helping in understanding if in

Table 1. Summary: we consider a typical thermal WIMP in the MSSM which annihilates mainly into heavy quarks or gauge bosons and compare different detection techniques. The first column indicates roughly what fraction of the energy released per annihilation goes into each channel, the second the ratio of the expected signal to the measured fluxes, the third whether an excess with respect to standard sources has been claimed based on current data, the fourth if a clean signature may arise in upcoming measurements and the last whether there is correlation or not between the different channels.

	energy fraction	exotic / detected flux (no clumps)	excess in data	clean signature	correlation
antiprotons	$\sim 5\%$	up to ~ 1	no	maybe	+
antideuterons	$\sim 5 \cdot 10^{-4}\%$	no data	no data	yes	+ +
positrons	$\sim 10\%$	up to $\sim 1/10$	maybe	maybe	+ / -
cont. γ-rays	$\sim 20 - 30\%$	up to $\sim 1/5$	maybe	maybe	+ / -
γ-ray lines	up to 0.1%	no data	no data	yes	-
synchr. rad.	?	larger than 1?	no	?	+ / -

the global picture of cosmic rays there is room, or even need, for a dark matter induced component, and if such component can be unambiguously singled out. The next generation of gamma-ray experiments, both ground- and space-based, e.g.[46,47,48], will map for the first time the gamma-ray sky in the energy range between few GeV to few hundreds of GeV where the spectral features typical for the dark matter induced gamma-ray emission (both the component with continuum energy spectrum and the gamma-line) might be found. Dark matter detection is not the main focus of these experiments, but such discovery may come out of data collected towards the Galactic center, or at high latitudes, or by mapping of the diffuse extragalactic gamma-ray background[18].

Table 1 offers a schematic summary of the points discussed in this review, enlightening common features and complementarities of the various techniques presented here. The answer one of the most pressing questions in astroparticle physics may indeed be around the corner and come from cosmic ray measurements.

References

1. M.S. Turner, *Astrophys. J. Lett.* in press, astro-ph/0106035.
2. G. Jungman, M. Kamionkowski and K. Griest, *Phys. Rep.* **267**, 195 (1996).
3. W. Hu *et al*, *Astrophys. J.* **549**, 669 (2001).
4. T. Gherghetta, G.F. Giudice and J.D. Wells, *Nucl. Phys.* **B559**, 27 (1999).
5. S. Cole, C.G. Lacey, C.M. Baugh and C.S. Frenk, *Mon. Not. R. Astron. Soc.* **319**, 168 (2000).
6. J.F. Navarro, C.S. Frenk and S.D.M. White, *Astrophys. J.* **462**, 563 (1996).
7. B. Moore *et al*, *Astrophys. J.* **499**, L5 (1998).
8. J. Silk and M. Srednicki, *Phys. Rev. Lett.* **50**, 624 (1984).
9. F.W. Stecker, S. Rudaz and T.F. Walsh, *Phys. Rev. Lett.* **55**, 2622 (1985).
10. K. Griest and D. Seckel, *Phys. Rev.* **D43**, 3191 (1991).
11. J. Edsjö and P. Gondolo, *Phys. Rev.* **D56**, 1879 (1997).
12. L. Bergström, J. Edsjö, P. Gondolo and P. Ullio *Phys. Rev.* **D59**, 043506 (1999).
13. P. Gondolo and J. Silk, *Phys. Rev. Lett.* **83**, 1719 (1999).

14. P. Ullio, H.S. Zhao and M. Kamionkowski, *Phys. Rev.* **D64**, 043504 (2001).
15. J. Silk, K. Olive and M. Srednicki, *Phys. Rev. Lett.* **55**, 257 (1985).
16. J.A. Tyson, G.P. Kochanski and I.P. Dell'Antonio *Astrophys. J. Lett.* **498**, L107 (1998).
17. E.A. Baltz et al, *Phys. Rev.* **D61**, 023514 (2000).
18. L. Bergström, J. Edsjö and P. Ullio, *Phys. Rev. Lett.* **87**, 251301 (2001).
19. F.Donato et al, *Astrophys. J.* **536**, 172 (2001).
20. I.V. Moskalenko, A.W. Strong, J.F. Ormes and M.S. Potgieter, *Astrophys. J.* **565**, 280 (2002).
21. S. Colafrancesco and B. Mele, *Astrophys. J.* **562**, 24 (2001).
22. P. Blasi, A.V. Olinto and C. Tyler, astro-ph/0202049.
23. A. Buffington et al, *Astrophys. J.* **248**, 1179 (1981).
24. T. Sanuki et al, *Astrophys. J.* **545**, 1135 (2000).
25. M. Boezio et al, *Astrophys. J.* **561**, 787 (2001).
26. P. Ullio, PhD thesis, Physics Department, Stockholm University, (1999).
27. F. Donato, N. Fornengo and P. Salati, *Phys. Rev.* **D62**, 043003 (2000).
28. K. Mori et al, *Astrophys. J.* in press, astro-ph/0109463.
29. M. Kamionkowski and M.S. Turner, *Phys. Rev.* **D43**, 1774 (1991).
30. E.A. Baltz and J. Edsjö, *Phys. Rev.* **D59**, 023511 (1999).
31. S. Ahlen et al, *Nucl. Instrum. Methods* **A350**, 351 (1994).
32. O. Adriani et al, *Proc. of the 26th ICRC*, Salt Lake City, OG.4.2.04 (1999).
33. J. Silk and H. Bloemen, *Astrophys. J.* **313**, L47 (1987).
34. T.K. Gaisser, *Cosmic Rays and Particle Physics*, Cambridge University Press, (1991).
35. L. Bergström and H. Snellman, *Phys. Rev.* **D37**, 3737 (1988).
36. L. Bergström and P. Ullio, *Nucl. Phys.* **B504**, 27 (1997).
37. P. Ullio and L. Bergström, *Phys. Rev.* **D57**, 1962 (1998).
38. P. Ullio, *JHEP* **0106**, 053 (2001).
39. S.W. Barwick et al, *Astrophys. J.* **482**, L191 (1997).
40. M. Boezio et al, *Astrophys. J.* **532**, 653 (2000).
41. G.L. Kane, L.T. Wang and J.D. Wells, *Phys. Rev.* **D65**, 057701 (2002). hep-ph/0108138.
42. H.A. Mayer-Hasselwander et al, *Astron. Astrophys.* **335**, 161 (1998).
43. D.D. Dixon et al, *New Astron.* **3**, 539 (1998).
44. A.W. Strong, I.V. Moskalenko and O. Reimer *Astrophys. J.* **537**, 763 (2000); *Erratum-ibid.* **541**, 1109 (2000).
45. L. Bergstrom, P. Ullio and J. Buckley, *Astropart. Phys.* **9**, 137 (1998).
46. M. Martinez et al, *Proc. of the 26th ICRC*, Salt Lake City, OG.4.3.08 (1999).
47. Gamma-ray Large Area Space Telescope (GLAST), homepage http://www-glast.stanford.edu.
48. R. Battiston et al., *Astropart. Phys.* **13**, 51 (2000).

INTEGRAL:
A GAMMA-RAY OBSERVATORY

NICOLAS PRODUIT

ISDC, INTEGRAL Science Data Center, Chemin d'cogia 16
Versoix, CH-1290,Switzerland

INTEGRAL[1] is an ESA mission designed to study the γ-ray sky in the 10 keV to 10 MeV energy band. It will build upon previous results and expand our knowledge of the γ-ray astronomy. It is organised as an observatory and thus can be exploited by a wide community.

Keywords: gamma-ray astronomy; INTEGRAL.

1. The INTEGRAL mission

INTEGRAL (INTErnational Gamma-Ray Astrophysics Laboratory) (see Fig. 1) is a satellite of the European Space agency (ESA) for low-energy γ-ray astronomy. It is a medium-size scientific mission of the *Horizon 2000*[2] program.

INTEGRAL will be launched on the 17th of October 2002 on board a Russian PROTON rocket in Baikonur. The orbit will be highly eccentric: 10'000 km perigee, 152'700 km apogee, 56.1°,72 sideral hours. This orbit is outside of the Van Allen belts most of the time, a very important consideration for detectors very sensitive to induced radioactivity. This orbit provides also very short eclipse seasons in which the spacecraft has to be put some time in a dormant state when not enough energy is provided by sunlight. The nominal lifetime of the observatory will be 2 years

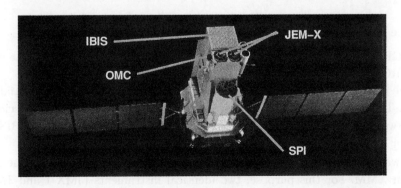

Fig. 1. The INTEGRAL spacecraft with its instruments identified

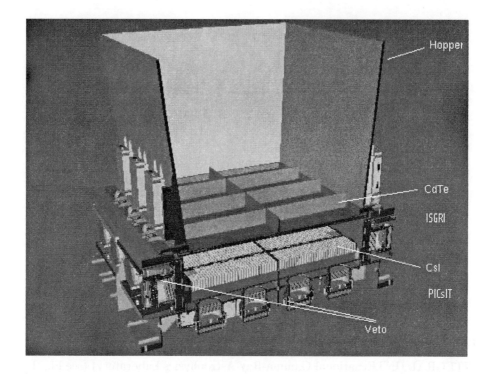

Fig. 2. The IBIS two sensitive planes, the active and the passive shielding

with possible extension up to 5 years. The mission utilizes the same service module (bus) developed for the ESA XMM[3] project. The spacecraft can be pointed in any direction of the sky except in the direction of the sun and the moon as well in the direction opposite to the sun (the solar panels are fixed perpendicular to the pointing direction and the moon is confusing to the star trackers). The pointing of the spacecraft is measured by two star trackers and maintained by reaction wheels and thrusters.

2. The instruments

INTEGRAL was designed by ESA in such a way that it will build upon the results of CGRO[4] and SIGMA[5] on board GRANAT.

The goal was to reach a high spectral resolution as well as a high spatial resolution. This was not possible to achieve in just one instrument so it was decided to build two main coalligned instruments, one mainly devoted to imaging (IBIS) and another one mainly devoted to spectroscopy (SPI).

It was also deemed important to insure some energy overlap with the X-ray mission XMM. For this purpose, two identical instruments (JMX-1 and JMX-2) cover the hard X-ray range up to the energy of the two main instruments. Those two instruments are also coalligned with the main instruments.

Fig. 3. The IBIS flight model mask during integration

Finally an optical camera (OMC) was added to provide for optical counterparts of the γ-ray sources. This optical measurement is of paramount importance when trying to identify new γ-rays sources.

2.1. IBIS

The IBIS (Imager on Board the INTEGRAL Satellite) telescope is a soft γ-ray (20 keV to 10 MeV) device based on a coded aperture imaging system (see Fig. 2 for the sensitive detector planes and Fig. 3 for the mask). In this energy range, conventional focusing techniques are difficult or even impossible to implement. The background due to natural and induced radioactivity is very high and variable. In coded aperture telescopes, source radiation is spatially modulated by a mask of opaque and transparent elements before being recorded on a position sensitive detector (see Fig. 4). This allows for simultaneous measurement of source and background fluxes. The mask has a mathematical property that enables the reconstruc-

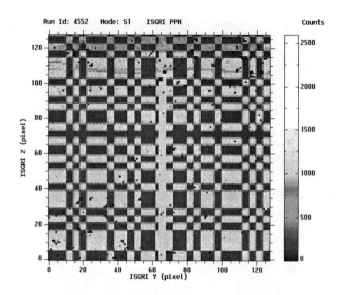

Fig. 4. A shadowgram. The image of the mask is projected on the sensitive plane

tion of the sky image based on the correlation of the recorded image and a constant array.[6]

The angular resolution of such a system is defined by the angle subtended by one hole of the mask at the detector plane. In IBIS the distance between the mask and the sensitive detector is 3.2 m.

There are two detector planes 10 cm appart: ISGRI above and PICsIT below. ISGRI[7] is a 128 × 128 pixels ZnCdTe detector array. PICsIT[9] is a 64 × 64 pixels CsI detector array. The two plane geometry enables the geometrical selection of Compton events and even the measurement of photon polarization. An anticoincidence shield is build around the sensitive planes to reject out-of-field-of-view photons as well as non-contained Compton events.

IBIS will achieve an angular resolution of 12' over an energy range between 15 keV and 10 MeV with an energy resolution of 10 keV @ 100 keV. The fully coded field of view is 9° × 9°.

2.2. SPI

SPI[8] (SPectrometer on INTEGRAL) is also composed of a mask, an anticoincidence shield and a sensitive plane. But here the mask is much more coarse, the anticoincidence much bigger and there are only 19 sensitive detectors. The detectors are made of high-purity germanium (see Fig. 5).

A cooling system will allow the spectrometer to reach a temperature of 85°K. Cosmic protons and neutrons will slowly damage the germanium crystals resulting in a loss of resolution and efficiency of the detector. To counteract this phenomena

Fig. 5. The SPI sensitive detectors plane

and restore the crystals, the detector will be occasionally heated up to a temperature of 100 °C for 24 hours.

The field of view will be 16 degrees with an angular resolution of 2.5 degrees. The energy range will be from 20 keV to 8 MeV with a resolution of 2.3 keV at 1.3 MeV.

2.3. JEM-X

There are two identical X-ray monitors, JEM-X (Joint European X-ray Monitor) 1 and 2, which complement the two previously mentioned instruments and which will play a decisive role in the detection and identification of γ-ray sources.

JEM-X will work simultaneously with the other instruments in the 3–35 keV energy range and with an angular resolution of three arcmin and a field of view of 4.8 °. Like the two previous instruments, the X-ray monitor will use a coded mask, in this case, placed 3.4 meters above the detector. The detector consists of two identical chambers (see Fig. 6) filled with Xenon at a pressure of 1.5 bar.

2.4. OMC

The optical monitoring camera (OMC) is composed of a CCD detector of 1024 × 1024 pixels located in the focal plane of a 50 mm lens with a V (visible) filter (see Fig. 7). Each pixel has a size of 13 × 13μm covering a field of 17.6 × 17.6 arcsec.

Fig. 6. The JEM-X gas chamber

The total field of view of the OMC camera will be of 5 × 5 degrees. The camera is placed on top of the satellite and is sensitive to stars of magnitude down to 19.7.

Combining the performances of all those detectors result in the sensitivity curve shown in Fig. 8.

3. Science with INTEGRAL

The atmosphere that shields the Earth from the γ-rays is responsible for the late development of this branch of astrophysics. The other problem of γ-ray astronomy lies in their very nature. With an equal amount of energy, a source would emit a million times less γ photons than visible photons.

Astrophysical sources of γ-rays are always extremely violent phenomena. Except for the planetary movements, the optical night sky remains identical night after night. γ-ray sources in the contrary seem to flare up and die continuously. Some sources only appear once in brief pulses and seem to disappear forever.

The main source of γ-rays are:

- The sun. But INTEGRAL is not devoted to and cannot study this source.
- Interacting binaries. Binary stars are pairs of stars sufficiently close to each other to be held together by their mutual gravitational attraction. Such pairs of stars

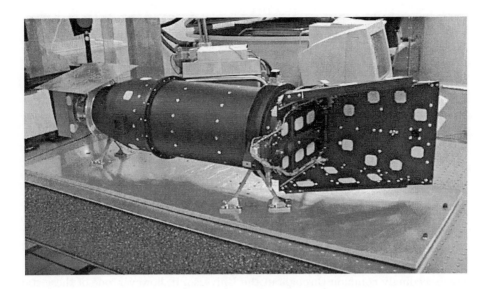

Fig. 7. The OMC camera

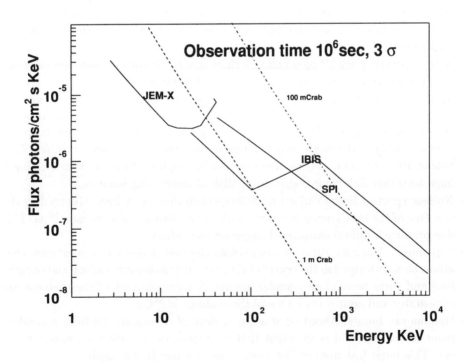

Fig. 8. Sensitivity of INTEGRAL for continuum photons

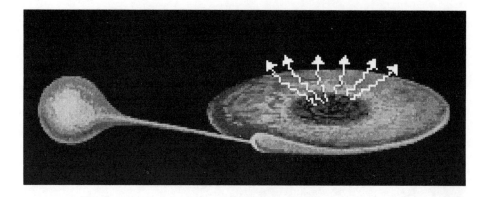

Fig. 9. Binary system showing an acretion disk

are exceedingly common throughout the Universe. If, however, one of these stars collapses and becomes a compact object, like a neutron star or a black hole, its huge gravitational field, will cause matter to be drawn from the surface of its companion star sweeping it around into an accretion disk (see Fig. 9). The matter of this disk spirals faster and faster towards the center and by dynamic friction reaches very high temperatures (typically 10^7K), thus emitting X-rays and γ-rays.

- Active galactic nuclei. Some galaxies show a very bright emission coming from their center, and are, therefore, called active galactic nuclei (AGN). There are different types of active nuclei, the most interesting of which for γ-ray astronomy are the Seyfert galaxies and quasars. The fact that they emit in all wavelengths of the electromagnetic spectrum implies highly complex mechanisms. The luminosity of quasars and Seyfert galaxies is attributed to the presence of a super-massive black hole of 10^8–10^{10} solar masses dragging nearby matter into them, thus forming a huge accretion disk comparable to the disk of interacting binaries.
- Nuclear spectral lines. Various nuclear spectral lines have been observed in the sky. One of the best known line is of a radioactive isotope of aluminum ^{26}Al. This line produce a diffuse emission throughout the galaxy.
- Supernovae. During a supernova explosion, the energy density is so extreme that atoms already formed in the center of the star, such as helium, carbon and oxygen, fuse and form new, more complex atoms. A large amount of these atoms are radioactive and emit γ-rays. One of these atoms is ^{56}Co.
- Gamma-ray bursts. About once a day a flare of γ-rays arrives from a random point in the sky, and is so bright that it outpowers any other γ sources in the sky. This burst last from milliseconds to several hundred seconds.
- Pulsars. Many pulsars do emit γ-rays. Some pulsars (Geminga) seem in fact to emit mostly in the γ-ray range and are quiet in the radio band. A pulsar's pulsed emission is very stable, more stable than the best atomic clocks.

In all those fields INTEGRAL will be able to make measurements of unsurpassed quality. For example INTEGRAL will be in a very unique position to perform fine spectroscopy and time evolution of gamma-ray bursts and be the first instrument sensitive to sub-millisecond gamma-ray burst. As the γ-ray sky is very unpredictable we do expect INTEGRAL to discover many new sources and perhaps new category of objects.

4. Concept of an observatory

INTEGRAL will be operated as an observatory. This means that there will be regular *Announcement of Opportunity* (AO) that will offer the opportunity to any astronomer to ask for specific observations from the facility. The *Integral Time Allocation Committee* will then judge the different proposals. If the demand is judged scientifically sound and feasible an observation will be scheduled during the *Open Time*. The Open Time consist of half the total observing time, the other half is the *Guaranteed Time* reserved to the member of the *Integral Science Working Team*, representing mostly laboratories which have contributed to the INTEGRAL program.

When an observation is completed, the ISDC[10] (INTEGRAL Science Data Center) processes all the data and provides the observer a set of predefined standard products: the raw data, the corrected and calibrated data, images, list of found sources, spectrum from each source... The ISDC also provides all the software used to produce the standard products as well as software that enables the Guest Observer to make further analysis of his data.

The Observer has the exclusive data rights on his observation for one year after he receives the standard products. After this period the ISDC makes this data available to anybody interested.

4.1. *The Open Time*

The first Announcement of Opportunity has already taken place and many observations have been granted. Some examples are:

- Probing core collapse: ^{44}Ti and ^{60}Co nucleosynthesis in SN 1987A
- View of Mkn 3 and Mkn 6: two Seyfert galaxies with strong absorption and radio jets
- Inverse Compton catastrophe and pair creation in the intraday-variable sources 0716+714 and 0836+71
- Exposing the binary heart of η Carinae with γ-rays
- Probing an intermediate mass Black Hole
- The physics of AGN: a deep understanding of the quasar 3C 273
- The Cygnus-X region: a nucleosynthesis laboratory
- To the bottom of the explosion forming Cas A: observing 44Ti and the hard X-ray emission

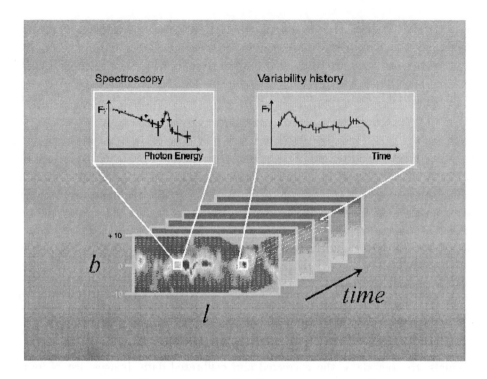

Fig. 10. Weekly scan of the galactic plane. b is the galactic latitude, l the galactic longitude.

- Deep view of the galactic center

The pressure factor (time requested over time available) reached a value of 20 which demonstrates that the community was very eager to receive this kind of data. Another reason is that spectroscopy and imaging of γ-rays is inherently a slow process, due to the very unfavorable ratio of real astrophysical photons to photons arising from the activated parts of the detector itself.

4.2. The Guaranteed Time

For the Guaranteed Time, the Integral Science Working Team has decided to perform the following observations:

- A Deep exposure of the Galactic Central Radian: $4.30 \ 10^6$ sec
- Weekly scans of the Galactic Plane (see Fig. 10) : $2.30 \ 10^6$ sec
- Pointed observations of the Vela region: $2.72 \ 10^6$ sec

The Vela and the Galactic Central Radian are extremently important regions that require so long exposures that cannot really fit in the guest observer program. The weekly scan will insure that no important transient signals are overlooked and it will trigger the accepted *Target Of Opportunity* observations. INTEGRAL will also accept important *Target Of Opportunity* observations triggered by external means.

5. Conclusions

With the very successfull XMM[3] and Chandra[11] missions devoted to X-rays and, in the future, the GLAST[12] and AMS[13,14] missions in space and the AUGER[15] observatory on earth devoted to very high energy γ-rays, INTEGRAL is the logical next step to close the energy gap.

This energy range is the most difficult to measure due to the fact that focalisation of γ-rays is not possible and signal over noise is very unfavorable.

INTEGRAL will feature detectors well capable of measuring the three (four) observables of photons: time of arrival, direction, energy (and polarization).

We look forward to the 17th October 2002 to start collecting exciting physics.

References

1. C. Winkler, *Proc. 3nd INTEGRAL Workshop - Astro. Lett. and Communications*, **39**, 309 (1999).
2. Horizon 2000 is the twenty-year programme formulated in 1984 by the European Space Agency and due to terminate in 2006.
3. Lumb, D. *et al.*, *Proceedings of the SPIE* v 2808, 326-337 (1996)
4. N. Gehrels and C.R. Shrader, *Astron. Astrophys. suppl. Ser.* **120**, 1-4 (1996)
5. Goldwurm, A., *Exp. Astron.*, 6, 9 (1995)
6. E. Caroli, *et al.*, *Space Sci. Rev.*, 45 , 349 (1987).
7. F. Lebrun *et al.*, *Nucl. Instr. & meth.* **A380**, 414 (1996).
8. Jean P., Vedrenne G., Schönfelder V., *et al.*, *Proceedings of the fifth Compton Symposium. AIP Conference Proceedings*, Vol. 510, p.708, (2000).
9. C. Lebanti *et al.*, *Proc. of SPIE Symposium.*
10. T. Courvoisier, *et al.*, *Proc. 3nd INTEGRAL Workshop - Astro. Lett. and Communications*, **39**, 355 (1999).
11. Weisskopf, M.C., *et al.*, PASP, 114, 791, 1-24 (2002)
12. Kniffen, D. A, *et al.*, *AIP Conference Proceedings*, Vol. 515., p.492 (2000)
13. S.Ahlen et al, NIM A350 (1994) 351.
14. R. Battiston, *these proceedings.*
15. Matthews, J., *American Astronomical Society Meeting 192*, 62.18 (1998)

THE AGILE MISSION AND GAMMA-RAY ASTROPHYSICS

MARCO TAVANI

Istituto di Astrofisica Spaziale e Fisica Cosmica, CNR
via Bassini 15, I-20233 Milano (Italy)

Gamma-ray astrophysics in the energy range between 30 MeV and 30 GeV is in desperate need of arcminute angular resolution and source monitoring capability. The AGILE Mission planned to be operational in 2004-2006 will be the only space mission entirely dedicated to gamma-ray astrophysics above 30 MeV. The main characteristics of AGILE are the simultaneous X-ray and gamma-ray imaging capability (reaching arcminute resolution) and excellent gamma-ray timing (10-100 microseconds). AGILE scientific program will emphasize a quick response to gamma-ray transients and multiwavelength studies of gamma-ray sources.

Keywords: Gamma-ray, astrophysics

1. Introduction

Forty years of study of the gamma-ray Universe produced remarkable results and many puzzles[10]. Several space missions contributed to the current view of the gamma-ray sky, including the third Orbiting Solar Observatory (OSO-3) active in the late Sixties[15], the SAS-2 satellite (Nov. 1972 – July 1973) [8], the European Cosmic-Ray Satellite[6,20] (COS-B, August 1975 – April 1982), and the Gamma-Ray Observatory (GRO) with the EGRET instrument[9,25] (April 1991 – June 2000). About 300 gamma-ray sources above 30 MeV were detected including about 70 active galactic nuclei (AGNs)[12,11,26,27], and seven isolated pulsars[28]. Interesting results were obtained on the cosmic-ray origin [18], diffuse Galactic[7] and extragalactic[19] emissions. At least five gamma-ray bursts (GRBs) were clearly detected and imaged above 30 MeV[7,13].

However, if we want to make further progress, we need a substantial improvement of the scientific performance of gamma-ray instruments. We are in desperate needs of:

- **better angular resolution and a larger field of view (FOV)** (an apparently contradictory request !), improving EGRET error boxes and FOV by a factor of at least 4;
- **better timing**, reaching a few microsecond for photon tagging and a deadtime for gamma-ray detection below 1 ms;
- **simultaneous X-ray and gamma-ray detection of unidentified sources**.

The AGILE Mission has these capabilities.

2. The AGILE Mission

The space program AGILE (*Astro-rivelatore Gamma a Immagini LEggero*) is an ASI Scientific Mission[22,23] entirely dedicated to gamma-ray astrophysics (30 MeV–50 GeV). AGILE is currently in Phase C[3], and is planned to be operational during the period 2004-2006. The AGILE scientific instrument is based on the state-of-the-art and reliably developed technology of solid state Silicon detectors developed in the Italian INFN laboratories[1,2,4]. The instrument is relatively light ($\sim$ 100 kg) and effective in detecting and monitoring gamma-ray sources in the energy range 30 MeV–50 GeV within a large field of view ($\sim$ 1/5 of the whole sky).

The AGILE instrument is made of three detectors with broad-band detection and imaging capabilities. The Gamma-Ray Imaging Detector (GRID) is sensitive in the energy range $\sim$ 30 MeV–50 GeV. It is characterized by the smallest ever obtained deadtime for single gamma-ray detection ($\lesssim$100 μs), and by a trigger logic based exclusively on Silicon plane detectors. The GRID consists of a Silicon-Tungsten Tracker, a Cesium Iodide Mini-Calorimeter, an Anticoincidence system made of segmented plastic scintillators, and fast readout electronics and processing units. The GRID is designed to achieve an optimal angular resolution (source location accuracy $\sim$ 5' – 20' for intense sources), an unprecedently large field-of-view ($\sim$ 3 sr), and a sensitivity comparable to that of EGRET for on-axis (and substantially better for off-axis) point sources.

AGILE will also have detection and imaging capabilities in the hard X-ray range provided by the Super-AGILE detector. It consists of an additional plane of four Silicon square detectors positioned on top of the GRID Tracker plus an ultra-light coded mask structure whose top absorbing mask at the distance of 14 cm from the silicon detectors. The main goals of Super-AGILE are the simultaneous gamma-ray and hard X-ray detection of astrophysical sources (unprecedented for gamma-ray instruments), optimal source positioning (1-3 arcmins, depending on intensity), fast burst alert and on-board trigger capability.

The CsI Mini-Calorimeter (MC) will also detect and collect events independently of the GRID. The energy range for this non-imaging detector is 0.3–200 MeV, and it can be very useful to provide spectral and accurate timing information of transient events.

AGILE with its combination of GRID, MC, and Super-AGILE is an innovative instrument, with an optimal expected performance for transients (AGNs, GRBs, stellar flares, unidentified gamma-ray sources) and steady sources (e.g., pulsars). The fast AGILE electronic readout and data processing (resulting in very small detectors' deadtimes) allow for the first time the systematic search for sub-millisecond gamma-ray transients with durations comparable with the dynamical timescale of $\sim 1\,M_\odot$ compact objects[24].

The AGILE Science Program will be focused on a prompt response to gamma-ray transients and alert for follow-up multiwavelength observations. AGILE will provide crucial information complementary to many space missions (INTEGRAL,

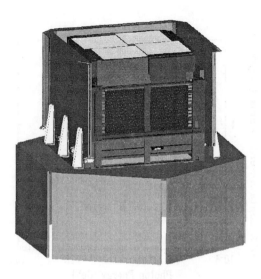

Fig. 1. Lateral view of the AGILE instrument (AC System and spacecraft partially displayed). The GRID is made of a Silicon Tracker (12 Tungsten and Silicon planes) and a Mini-Calorimeter is placed at the bottom of the istrument. Super-AGILE has its 4 Si-detectors placed at the top of the first GRID tray, and an ultra-light coded mask system (CMS) positioned on top (the figure shows the CMS partition configuration). The instrument size is $\sim 63 \times 63 \times 58.5\,cm^3$, including Super-AGILE and the AC System for a total weight of ~ 100 kg.

XMM, CHANDRA, SWIFT, and others). Furthermore, it can support ground-based investigations in the radio, optical, and TeV bands. Quicklook data analysis and fast communication of new transients will be implemented as an essential part of the AGILE Science Program. No other mission entirely dedicated to gamma-ray astrophysics above 30 MeV is being planned before GLAST.

3. Science with AGILE

3.1. Gamma-Ray Astrophysics with the GRID

The GRID is designed to obtain:

- **excellent imaging capability** in the energy range 100 MeV-50 GeV, improving the EGRET error box radius by a factor of 2 (see Fig. 2);
- **a very large field-of-view**, allowing simultaneous coverage of $\sim 1/5$ of the entire sky per each pointing (FOV larger by a factor of ~ 5 than that of EGRET) with an effective area at 400 MeV of $A_{eff} \simeq 550\ cm^2$ on-axis, and $A_{eff} \simeq 350\ cm^2$ for inclination angles of 50 degrees off-axis;
- **excellent timing capability**, with absolute time tagging of uncertainty near 1 μs and very small deadtimes ($\sim 100\,\mu$s for the Si-Tracker and $\sim 20\,\mu$s for each of the individual CsI bars);

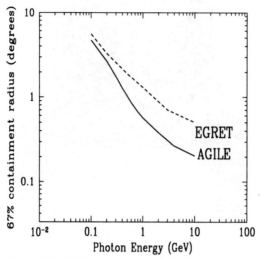

Fig. 2. Three dimensional PSF (67% containment radius) as a function of photon energy for AGILE-GRID and EGRET.

- **a good sensitivity for point sources**, comparable to that of EGRET for *on-axis* sources, and substantially better for *off-axis* sources;
- **excellent sensitivity to photons in the energy range** ~30-100 MeV, with an effective area above 200 cm^2 at 50 MeV;
- **a very rapid response to gamma-ray transients and gamma-ray bursts**, obtained by a special quicklook analysis program and coordinated ground-based and space observations.

3.1.1. *Large FOV monitoring of gamma-ray sources*

Fig. 3 show a typical AGILE pointing. Relatively bright AGNs and Galactic sources flaring with fluxes larger than 10^{-6} ph cm^{-2} s^{-1} (above 100 MeV) can be detected within a few days by the AGILE quicklook analysis. We conservatively estimate that for a 3-year mission AGILE is potentially able to detect a number of gamma-ray flaring AGNs larger by a factor of several compared to that obtained by EGRET during its 6-year mission. Furthermore, the large FOV will favor the detection of fast transients such as gamma-ray bursts. Taking into account the high-energy distribution of GRB emission above 30 MeV, we conservatively estimate that ~1 GRB/month can be detected and imaged in the gamma-ray range by the GRID.

3.1.2. *Fast reaction to strong high-energy transients*

The existence of a large number of variable gamma-ray sources (extragalactic and near the Galactic plane[21]) makes necessary a reliable program for quick response to

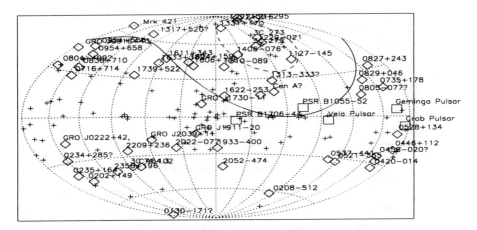

Fig. 3. Comparison between a typical AGILE pointing centered at the blazar 3C 279 (GRID FOV within the solid line circle of radius equal to 60°) and by EGRET (FOV within the dashed line of radius equal to 25°).

transient gamma-ray emission. Quicklook analysis of gamma-ray data is a crucial task to be carried out by the AGILE Team. Prompt communication of gamma-ray transients (that require typically 2-3 days to be detected with high confidence for sources above 10^{-6} ph cm^{-2} s^{-1}) will be ensured. Detection of short timescale (seconds/minutes/hours) transients (GRBs, SGRs, solar flares and other bursting events) is possible in the gamma-ray range. A primary responsibility of the AGILE Team will be to provide positioning of short-timescale transient as accurate as possible, and to alert the community though dedicated channels.

3.1.3. Large exposures for Galactic and extragalactic sky regions

The AGILE average exposure per source will be larger by a factor of ~ 4 for a 1-year sky-survey program compared to the typical exposure obtainable by EGRET for the same time period. After a 1-year all-sky pointing program, AGILE average sensitivity to a generic gamma-ray source above the Galactic plane is expected to be better than EGRET by a factor conservatively given as ~ 2. Deep exposures for selected regions of the sky can be obtained by a proper program with repeated overlapping pointings. For selected regions, AGILE can achieve a sensitivity larger than EGRET by a factor of $\sim 4-5$ at the completion of its program. This can be particularly useful to study selected Galactic and extragalactic sources. For selected sky areas, AGILE can then achieve a flux sensitivity better than 5×10^{-8} ph cm^{-2} s^{-1} at the completion of its scientific program.

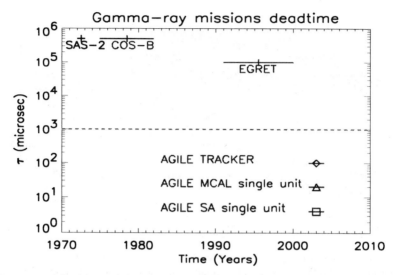

Fig. 4. Instumental deadtimes of the AGILE detectors and of previous gamma-ray instruments.

3.1.4. High-Precision Timing

AGILE detectors will have optimal timing capabilities. The on-board GPS system allows to reach an absolute time tagging precision for individual photons near 2 μs. Depending on the detectors hardware and electronics, absolute time tagging can achieve values near $1 - 2\,\mu$s for the Silicon-tracker, and $3 - 4\,\mu$s for the individual detecting units of the Mini-Calorimeter and Super-AGILE.

Instrumental deadtimes will be unprecedently small for gamma-ray detection. The GRID deadtime will be of order of 10 μs (improving by three orders of magnitude the performance of previous spark-chamber detectors such as EGRET). The deadtime of MC single CsI bars is near 20 μs, and that of single Super-AGILE readout units is $\sim 5\,\mu$s. Taking into account the segmentation of the electronic readout of MC and Super-AGILE detectors (30 MC elements and 16 Super-AGILE elements) the effective deadtimes will be much less than those for individual units.

Fig. 4 show the AGILE deadtime performance compared to other gamma-ray missions. Fast AGILE timing will, for the first time, allow investigations and searches for sub-millisecond transients in the gamma-ray energy range.

3.2. Super-AGILE

An imaging coded mask detector system (Super-AGILE) in addition to the GRID will provide a unique tool for the study of high-energy sources. The Super-AGILE FOV is expected to be ~ 0.8 sr. Super-AGILE can provide important information including:

- **source detection and spectral information in the energy range** $\sim$**10-40 keV** to be obtained simultaneously with gamma-ray data (5 mCrab

sensitivity at 15 keV (5σ) for a 50 ksec integration time);
- **accurate localization (~1-2 arcmins) of GRBs and other transient events** (for typical transient fluxes above ~1 Crab); the expected GRB detection rate is ~ 1 − 2 per month;
- **excellent timing**, with absolute time tagging uncertainty and deadtime near 4 μs for each of the 16 independent readout units of the Super-AGILE Si-detector;
- **long-timescale monitoring (~2 weeks) of hard X-ray sources**;
- **hard X-ray response to gamma-ray transients detected by the GRID**, obtainable by slight repointings of the AGILE spacecraft (if necessary) to include the gamma-ray flaring source in the Super-AGILE FOV.

The combination of simultaneous hard X-ray and gamma-ray data will provide a formidable combination for the study of high-energy sources. Given the GRID and Super-AGILE sensitivities, for the first time simultaneous hard X-ray/gamma-ray information will be obtainable for: (1) GRBs, (2) blazars with detectable hard X-ray emission such as 3C 273 and Mk 501, (3) Galactic jet-sources with favorable geometries, (4) unidentified variable gamma-ray sources. Table 1 summarizes the main characteristics of the AGILE detectors.

Table 1. AGILE Scientific Performance

Gamma-ray Imaging Detector (GRID)	
Energy Range	30 MeV – 50 GeV
Field of view	~ 3 sr
Sensitivity at 100 MeV (ph cm^{-2} s^{-1} MeV^{-1}, 5σ in 10^6 s)	6×10^{-9}
Sensitivity at 1 GeV (ph cm^{-2} s^{-1} MeV^{-1}, 5σ in 10^6 s)	4×10^{-11}
Angular Resolution at 1 GeV (68% cont. radius)	36 arcmin
Source Location Accuracy (S/N~10)	~5–20 arcmin
Energy Resolution (at 300 MeV)	$\Delta E/E \sim 1$
Absolute Time Resolution	~ 1 μs
Deadtime	~ 100 μs
Hard X–ray Imaging Detector (Super-AGILE)	
Energy Range	10 – 40 keV
Field of view (FW at Zero Sens.)	107°×68°
Sensitivity (at 15 keV, 5σ in 1 day)	~5 mCrab
Angular Resolution (pixel size)	~ 6 arcmin
Source Location Accuracy (S/N~10)	~2-3 arcmin
Energy Resolution	$\Delta E<4$ keV
Absolute Time Resolution	~ 4 μs
Deadtime (for each of the 16 readout units)	~ 4 μs
Mini-Calorimeter	
Energy Range	0.3 – 200 MeV
Energy Resolution (above 1 MeV)	~ 1 MeV
Absolute Time Resolution	~ 3 μs
Deadtime (for each of the 32 CsI bars)	~ 20 μs

3.3. Science with AGILE

We summarize here the main AGILE's scientific objectives.

- **Active Galactic Nuclei.** The large GRID field of view makes possible the simultaneous monitoring of a large number of AGNs per pointing. Several outstanding issues concerning the mechanism of AGN gamma-ray production and activity can be addressed by AGILE including: (1) the study of transient vs. low-level gamma-ray emission and AGN duty-cycles; (2) the relationship between the gamma-ray variability and the radio/optical/X-ray/TeV emission; (3) the correlation between major relativistic radio plasmoid ejections and gamma-ray flares; (4) simultaneous hard X-ray and gamma-ray emission properties. On the average, AGILE will achieve deep exposures of AGNs and substantially improve our knowledge on the low-level emission as well as detecting gamma-ray flares. We conservatively estimate that for a 3-year program AGILE will detect ~100 AGNs more than those already discovered by EGRET.

- **Gamma-ray bursts.** Five GRBs were detected by the EGRET spark chamber during 7 years of operations[16]. This number was limited only by the EGRET FOV and sensitivity and not by the GRB emission mechanism. Indeed, all GRBs occurring in the EGRET FOV and intense enough to be detectable were recorded by the spark chamber. The AGILE-GRID detection rate is expected to be ≥ 5 GRBs/year). The small GRID deadtime ($\sim$ 1000 times smaller than that of EGRET) allows to address the crucial issue of the impulsive particle acceleration and radiation during the initial phase of GRB pulses (for which EGRET response was in many cases inadequate for its deadtime of 100 ms). Furthermore, the remarkable discovery of 'delayed' gamma-ray emission up to $\sim$ 20 GeV from GRB 940217[13] opens the way to study gamma-ray emission on timescales much longer than the initial impulsive event. The GRID improved angular resolution for gamma-ray detection compared to that of EGRET allows a systematic search for delayed gamma-ray emission from GRBs. AGILE is also expected to be efficient in detecting photons above 10 GeV because of limited backsplashing even at large incidence angles. Super-AGILE will be able to locate GRBs within a few arcminutes, and will systematically study the interplay between hard X-ray and gamma-ray emissions. Special emphasis will be given to fast GRB timing search allowing the detection of sub-millisecond GRB pulses independently detectable by the Si-Tracker, MC and Super-AGILE.

- **Diffuse Galactic and extragalactic emission.** The AGILE good angular resolution and large average exposure will further improve our knowledge of cosmic ray origin, propagation, interaction and emission processes. We also note that a joint study of gamma-ray emission from MeV to TeV energies is possible by special programs involving AGILE and new-generation TeV observatories of improved angular resolution.

- **Gamma-ray pulsars.** AGILE will contribute to the study of gamma-ray pulsars in several ways: (1) improving photon statistics for gamma-ray period searches; (2) detecting possible secular fluctuations of the gamma-ray emission from neutron star magnetospheres; (3) studying unpulsed gamma-ray emission from plerions in supernova remnants and searching for time variability of pulsar wind/nebula interactions, e.g., as in the Crab nebula.

- **Search for non-blazar gamma-ray variable sources in the Galactic plane**, currently a new class of unidentified gamma-ray sources (GRO J1838-04[21]).
- **Galactic sources, new transients**. A large number of gamma-ray sources near the Galactic plane are unidentified, and sources such as 2CG 135+1 can be monitored on timescales of months/years. Also Galactic X-ray jet sources (such as Cyg X-3, GRS 1915+10, GRO J1655-40 and others) can produce detectable gamma-ray emission for favorable jet geometries, and a TOO program is planned to follow-up new discoveries of *micro-quasars*.

References

1. A. Bakaldin, A. Morselli, P. Picozza et al.,*Astroparticle Physics* **8**, 109 (1997).
2. G. Barbiellini et al., *Nuclear Physics B*, **43**, 253 (1995)
3. G. Barbiellini, G. et al.,in *Proceedings of the Symposium Gamma-2001*, AIP Conf. Proceedings, eds. S. Ritz, N. Gehrels, C.R. Shrader, Vol. 587, pp.774 (2001) 7
4. G. Barbiellini et al.., G. Bordignon, G. Fedel, M. Prest et al.NIM, in press (2002)
5. C.L. Bhat et al., *Nature*, **359**, 217 (1992)
6. G.F. Bignami and W. Hermsen,*Ann. Rev. Astr. Astrophys.*, **21**, 67 (1993)
7. B.L. Dingus et al., in *Second Huntsville Gamma-Ray Burst Workshop*, eds. G.J. Fishman, J.J. Brainerd and Hurley, K., AIP Conf. Proceeding no. 307, p. 22 (1994)
8. C.E. Fichtel et al., *ApJ*, **198**, 163 (1975).
9. C.E. Fichtel, *ApJS*, **94**, 551 (1994).
10. C.E. Fichtel and J.I. Trombka, *Gamma-Ray Astrophysics*, NASA Reference Publication No. 1386, Second Edition (1997)
11. R.C. Hartman, W. Collmar, W., C. von Montigny and C.D. Dermer, in *Proceedings of the Fourth Compton Symposium*, AIP Conf. Proceeding no. 401, p. 307 (1997)
12. R.C. Hartman et al., *ApJS*, 123, 279 (1999). (http://cossc.gsfc.nasa.gov/cossc/egret/3rd-EGRET-Cat.html)
13. K. Hurley et al.*Nature*, **372**, 652 (1994)
14. K. Koyama, et al.*Nature*, **378**, 255 (1995)
15. Kraushaar, W.L., et al., *ApJ*, **177**, 341 (1972)
16. Schneid E.J. et al., in AIP Conf. Proc. no. 384, p.253. (1996)
17. P. Sreekumar and C.E. Fichtel, *A&A*, **251**, 447 (1991)
18. P. Sreekumar et al., *Phys. Rev. Letters*, **70**, 127 (1993)
19. P. Sreekumar et al., *ApJ* **461**, 872 (1998)
20. B.N. Swanenburg et al., *ApJ*, **461**, 872 (1981)
21. M. Tavani et al., *ApJ*,**479**, L109 (1997)
22. M. Tavani, et.al., Proceedings of the 5th Compton Symposium, AIP Conf. Proceedings, ed. M. McConnell, Vol. 510, p. 746 (2002)
23. M. Tavani, et.al., Proceedings of the Symposium *Gamma-2001*, AIP Conf. Proceedings, eds. S. Ritz, N. Gehrels, C.R. Shrader, Vol. 587, p. 729 (2001)
24. M. Tavani, in preparation (2002)
25. D.J. Thompson et al.., *ApJS*, **86**, 629 (1993)
26. D.J. Thompson et al., *ApJS*, **101**, 259 (1995)
27. D.J. Thompson et al., *ApJS*, **107**, 227 (1996)

28. D.J. Thompson, A.K. Harding, W. Hermsen and M.P. Ulmer, in Proc. 4th CGRO Symp., eds. C.D. Dermer, M.S. Strickman, and J.D. Kurfess (New York, AIP Conf. Proc. no. 410), p. 39 (1998)

PULSARS, BLAZARS AND DARK MATTER WITH AMS

MARTIN POHL

DPNC, Université de Genève, 24 quai Ernest Ansermet
CH-1211 Genève 4, Switzerland

The detection potential of AMS for photon sources of different origin is assessed. In particular, galactic and extragalactic point sources detected by previous experiments are examined for their detectability at energies above a few GeV. In addition, gamma ray rates from a potential supersymmetric dark matter halo are estimated using different models. It is argued that the constant survey of the gamma ray sky by AMS with a high angular resolution will be an important complement to similar studies using other present and future instruments.

Keywords: Blazar; Pulsar; dark matter; AMS

1. Photon detection with AMS

High energy photons are detected in AMS either by calorimetric means or by conversion in the material above the spectrometer. In the first case the signature will be an electromagnetic shower in the lead-scintillator calorimeter at the bottom of the set-up. In the latter case, a pair of electron-positron tracks will be observed in the tracking detector as shown in Fig. 1.

For calorimetrically detected photons, the relative energy resolution is about 5% at 10 GeV and better than 3% at high energies above 100 GeV. The angular resolution for these photons is modest, about 1.5° at 10 GeV, reaching 0.5° at high energies[1].

Conversely, for converted photons[2] (see Fig. 2) the energy resolution is modest, a constant 15% below 100 GeV and slowly increasing above. The angular resolution is however excellent going down like a power law from about 0.6° at 1 GeV to 0.12 mrad at high energies[1].

In fact, a handful of photons has already been observed by conversion during the AMS-01 test flight in 1998[3]. Fig. 3 shows the energy distribution of the selected low mass opposite-sign electron pairs. The rate is in qualitative agreement with the expected photon rate, given a large and not very well known reduction by the trigger conditions during the test flight.

The effective aperture of AMS-02 for converted photons and for calorimetrically detected photons above 10 GeV is of the order of $0.1 m^2 sr$ each. Since both are independent instruments for the purpose of photon studies, their aperture is added in rate estimates below unless noted otherwise.

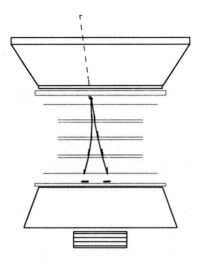

Fig. 1. Monte Carlo simulation of a converted photon seen in the AMS-02 tracking detector.

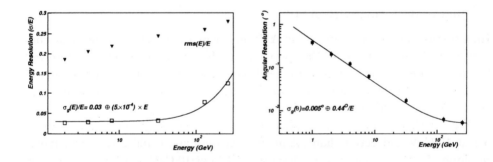

Fig. 2. Expected energy resolution (left) and angular resolution (right) as a function of energy for converted photons measured in the AMS-02 spectrometer.

One of the major differences between AMS-02 on the ISS and other photon detectors is that AMS will constantly scan large portions of the gamma ray sky over an extended period of time. This leads to an effective *area* × *time* product which is of the order of 100 cm^2 year inside the field of view, both for the tracking detector and for the calorimeter[4]. Figure 4 shows this rate determining quantity as a function of galactic coordinates for the calorimeter only.

The gamma ray sky between 100 MeV and 30 GeV has been extensively surveyed by the EGRET instrument on CGRO[5]. Among the identified sources are pulsars from the Milky Way as well as Active Galactic Nuclei from outside our galaxy.

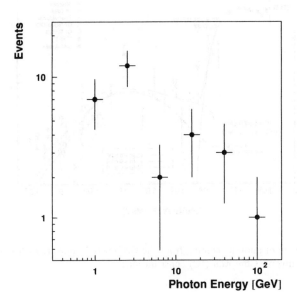

Fig. 3. Total energy distribution of selected low mass electron-positron pairs observed during the AMS-01 mission.

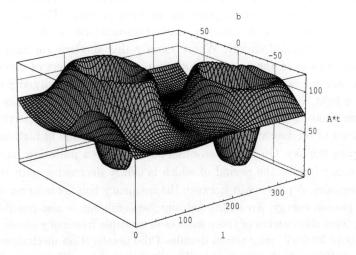

Fig. 4. Total *area* × *time* product in units of cm^{2}year for calorimetric photon detection during the three year AMS-02 mission on the ISS, as a function of galactic coordinates.

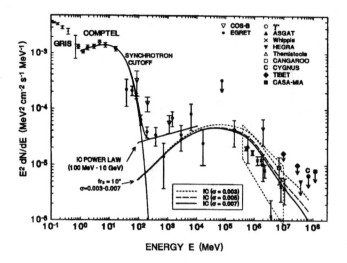

Fig. 5. Energy spectrum of photons coming from the Crab nebula pulsar, with model results for synchrotron light and inverse Compton (IC) contributions.

2. Galactic point sources

Pulsars are the most luminous galactic sources of high energy photons[6]. They are rotating neutron stars from supernovae remnants[7], for which the axis of rotation is misaligned with the magnetic axis. The large fields at the surface caused by this misalignment liberate electrons, positrons or even protons. The magnetosphere, dragged along by the rotation, can then channel the acceleration of charged particles fuelled by the rotational energy. Where exactly this acceleration happens, along the magnetic axis or in the so-called outer gap, is still a subject of debate. In any case, high energy photons are produced by the accelerated charged particles via synchrotron light. Inverse Compton scattering can then upgrade the energy of the synchrotron photons by scattering off further high energy charged particles.

This chain of mechanisms then leads to a collimated beam of high energy photons sweeping the sky. On Earth, an observer thus sees a pulsating photon signal in all frequency bands, the period of which is slowly decreasing with time as the rotation degrades. A phase shift between the frequency bands can occur which may vary with photon energy. An additional unpulsed emission is also possible.

A long term observation of these sources in multiple frequency bands, including energies above 30 GeV, may reveal details of the acceleration mechanism involved in pulsars. Different high energy cut-offs are predicted by different models. As an example, Fig. 5 shows the spectrum from the Crab nebula[8], remnant of a very young Supernova from the year 1054. The region from a few GeV to several hundred GeV is sparsely covered by experiments, yet important to fix the shape of the inverse Compton scattering contribution beyond the synchrotron radiation cut-off.

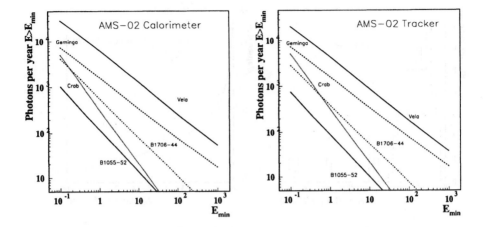

Fig. 6. Estimated photon rates from the most luminous pulsar sources for the calorimetric detection (left) and detection by conversion (right), as a function of the minimum detectable energy in GeV. The rates assume a continued power law behaviour of the spectrum at high energies.

To observe such signals, they must be distinguished from the diffuse galactic background[9] of several 10^{-8} photons/cm^2/sr/s/MeV observed above 1 GeV. The excellent angular resolution of especially the converted photons will ensure a clean measurement. Fig. 6 shows the expected photon rate per year above a given energy threshold if no natural high energy cut-off is present. From the most luminous sources, several hundreds to several thousands of photons will be registered above detection threshold by both detection methods. This estimate makes use of a rather accurate orbit and attitude model of the ISS, but only a rough estimate of the detector's reconstruction efficiency. In any case, one can conclude that there will be sufficient rate for a regular observation of the signal from these known sources, with an enlarged energy interval should the emission continue into high energies. These data will thus complement those gathered with other instruments at lower energies.

3. Extragalactic point sources

Among the most spectacular astrophysical objects are the Active Galactic Nuclei, AGN, of which there is a whole zoology[10]. They are believed to have a very massive rotating black hole in their center, surrounded by its accretion disk and a giant torus of gas and dust. Jets of particles are ejected along the rotation axis, fuelled by accretion and/or the rotational energy and confined through the magnetic fields generated by the rotating matter. Particle acceleration in these jets can happen by shock waves, other accelerating mechanisms are not excluded. Photons are emitted by synchrotron radiation and then boosted in energy by Compton scattering off

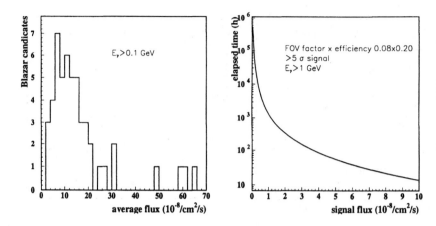

Fig. 7. Left: Flux distribution of blazars observed by EGRET above 100 MeV photon energy. Right: Estimate of the elapsed time required with AMS to see a 5σ significance signal from a source of a given flux above 1 GeV.

energetic charged particles. A contribution of hadronic jets with neutral pions is conceivable, the relative contribution of hadrons and electrons is under debate.

The observational characteristics and classification of AGN vary with the angle between the line of site and the jet axis. When the jet points towards Earth, the resulting violent object is called a blazar. Their photon emission rate is highly variable, while the spectral index stays constant[11]. These signals must be distinguished from the diffuse extragalactic background, three orders of magnitude weaker than the galactic one, but characterised by the same spectral index as the blazar signal[12]. Assuming a total efficiency of order 20% for detecting a photon and an efficient suppression of non-photon background[1], the rough sensitivity estimate shown in Fig. 7 is obtained. One concludes that all blazar sources in the EGRET catalogue[5] will be observable by AMS with high significance after elapsed times that vary from a month to a few hours, depending on the source's flux. With this increased sensitivity and improved angular resolution, the detection and identification of many new sources is likely.

4. Dark Matter halo of the Milky Way

The observed rotation curves of stars on the outer rim of many galaxies show that a large amount of matter, far beyond what can be accounted for by ordinary astrophysical objects, is present as a dark matter halo. An excellent candidate for being a main constituent of dark matter is the lightest supersymmetric particle[13]. In the Minimal Supersymmetric Standard Model (MSSM) and other models with R-parity conservation, this is usually the stable neutralino $\tilde{\chi}^0$, a neutral scalar boson which is its own antiparticle. Neutralino annihilation can thus lead to observable

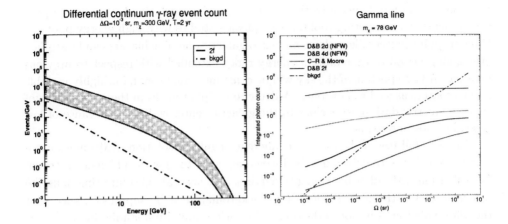

Fig. 8. Left: Contribution to the diffuse photon spectrum from the Milky Way's center in density distribution model 2f from Dehnen and Binney, for $m_{\tilde{\chi}} = 300$GeV. The width of the band indicates uncertainties in the annihilation cross section. Right: Integrated number of photons over a two year period from an annihilation line, $\tilde{\chi}^0\tilde{\chi}^0 \to \gamma\gamma$ with $m_{\tilde{\chi}} = 78$GeV, as a function of the angular resolution.

signals of light antiparticles ($\bar{p}$, e^+) and/or photons. Both annihilation lines from $\tilde{\chi}^0\tilde{\chi}^0 \to \gamma\gamma$ and contributions to the continuum from hadronic final states are conceivable. For high neutralino masses, the continuum contribution dominates. For modest masses, the annihilation cross section into photon pairs can also be sizable.

The density distribution of dark matter around galaxies is not known. A large class of models predicts the parameters α, β, γ, a and ρ_a of a generic density distribution

$$\rho(r) = \rho_a \left(\frac{r}{a}\right)^{-\gamma} \left[1 + \left(\frac{r}{a}\right)^{\alpha}\right]^{\frac{\gamma-\beta}{\alpha}} \qquad (1)$$

as a function of the distance r from the galactic center. A set of such models has been studied[4] to assess the detectability of continuum photons or an annihilation line. The most promising density distribution is the model 2f of Dehnen and Binney[14] with $\gamma \simeq 2$. For this model, the predicted continuum spectrum for $m_{\tilde{\chi}} = 300$GeV and the annihilation rate for $m_{\tilde{\chi}} = 78$GeV are shown in Fig. 8. For other models, the signals are weaker and can be undetectable if the density distribution is sufficiently diluted. Nevertheless, there is a finite chance of detecting a structure in the diffuse galactic background or a handful of events in a sharp annihilation line during a multi-year mission.

5. Conclusions

Before the advent of ultimate photon instruments like GLAST[15], photons can be detected with AMS using two complementary methods, with reasonable efficiency and good resolution. Especially, the excellent angular resolution for converted photons provides an interesting tool for astrophysics studies with high energy photons. Galactic point sources like pulsars and extragalactic ones like blazars can be studied over a long period of time and an energy range extended with respect to previous missions. A large portion of the sky will be constantly monitored, for highly variable known sources as well as new ones. A detection of photons from the annihilation of supersymmetric dark matter close to the galactic center is difficult because of low rates, but not impossible.

To make full use of the excellent inherent angular resolution, AMS must use a star tracker device to determine the instrument's attitude with sufficient precision. Coordination with other instruments requires that accurate absolute time must be provided. Non-astrophysical sources of background must be suppressed such that the diffuse photon radiation is the dominant background. First studies indicate that all of this should be feasible.

Acknowledgements

I wish to thank the organisers of this workshop for creating a forum of fruitful and friendly discussions. I am also grateful to my collaborators in AMS for providing much of the material in this talk prior to publication.

References

1. For updated performance estimates of the AMS detector, see V. Choutko, these proceedings.
2. B. Bertucci et al., *AMS-γ: high energy photons detection with the Alpha Magnetic Spectrometer on the ISS*, Proceedings of the 27th International Cosmic Ray Conference, Hamburg (2001), p. 2777
3. M. Cristinziani, *Search for Heavy Antimatter and Energetic Photons in Cosmic Rays with the AMS-01 Detector in Space*, PhD Thesis Université de Genève, 2002.
4. G. Valle, *Gamma Rates in AMS02 from Neutralino Annihilation*, AMS Internal Report 2001-11-04, November 2001.
5. R.C. Hartman et al., Astrophys. J. Suppl. **123** (1999) 79.
6. G. Kanbach et al., Astron. Astrophys. **289** (1994) 855;
   D.J. Thomson et al., Astrophys. J. **465** (1996) 385.
7. B.W. Caroll and D.A. Ostlie, *An Introduction to Modern Astrophysics*, Addison-Wesley Publishing Comp. (1996).
8. O.C. de Jager et al., Astrophys. J. **457** (1996) 253.
9. S.D. Hunter et al., Astrophys. J. **481** (1997) 205.
10. C.M. Urry and P. Padovani, Publ. Astron. Soc. Pacif. **107** (1995) 803.
11. R. Mukherjee et al., Astrophys. J. **490** (1997) 116.
12. P. Sreekumar et al., Astrophys. J. **494** (1998) 523.
13. L. Bergström, P. Ullio and J.H. Buckley, Astropart. Phys. **9** (1998) 137.
14. W. Dehnen and J. Binney, Mon. Not. R. Astron. Soc. **294** (1998) 429.
15. See e.g. H.F.W. Sadrozinski, Nucl. Instrum. Meth. **A466** (2001) 292.

COSMIC PHOTON AND POSITRON SPECTRA MEASUREMENTS MODELLING WITH THE AMS-02 DETECTOR AT ISS

VITALI CHOUTKO*

Laboratory For Nuclear Science, MIT, 77 Massachusetts Av., Cambridge MA 02171-9131, USA

GIOVANNI LAMANNA[†]

DPNC, Université de Genève, 24, quai E. Ansermet
1211 Geneva 4, Switzerland

ALEXANDER MALININ

Institute For Physical Science and Tech., UMD, Space Sci Bldg.,
College Park, MD 20742-0001, USA

The results of the MC feasibility study of the AMS-02 Detector cosmic-ray γ and e^+ spectra measurement capabilities are presented. AMS will be able to provide accurate measurements of the above spectra in a broad energy range (up to hundreds of GeV) with energy resolution of a few percent and angular resolution of 0.01-1°. The acceptance of 0.05-0.15 m²sr has been estimated both for e^+ and γ, with maximal opening angle being 20° to 45°. The systematic study of the background events has been done, the irreducible background to signal ratio below a few percent being obtained.

1. Introduction

The Alpha Magnetic Spectrometer (AMS)[1] is a high energy physics experiment scheduled for a three year mission on board the International Space Station.

The physics goals of the AMS are to search for antimatter in the Universe on the level of less than 10^{-9}, to search for dark matter and to make high statistics measurements of cosmic rays composition, as well as γ ray measurements over a broad energy range.

The preliminary version of the AMS experiment (AMS-01) recently had 100 hours engineering flight STS–91 on board of the space shuttle Discovery[2].

In this note we analysed the AMS capabilities of measuring of γ rays, which are crucial for the identification of various gamma sources in the Universe as well as for the precise measurement of the diffuse γ-ray emission and for the e^+ spectra measurement, the latter being considered as a favorite signature of the WIMP's detection, which may represent the bulk of the dark matter[3] in our Galaxy.

---

*E-mail:v.choutko@cern.ch

†Present address: CERN, European Organisation for Nuclear Research, 1211 Geneva 23, Switzerland

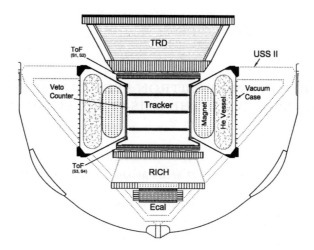

Fig. 1. Schematic view of the AMS-02 detector.

2. The AMS-02 Detector Simulation

The AMS-02 Detector geometry and performance was simulated with the dedicated AMS simulation and reconstruction program based on GEANT package[4].

The major elements of AMS-02 Detector are shown in Fig. 1 and consist of a superconducting magnet, a gaseous transition radiation detector (TRD), a silicon tracker (Tracker), time of flight hodoscopes (ToF), a ring imaging Cerenkov detector (RICH), an electromagnetic calorimeter (ECAL) and anticoincidence counters.

The superconducting magnet has the shape of a cylindrical shell with inner diameter 1.2 m and length 0.8 m and provides a central dipole field of 0.8 Tesla across the magnet bore.

The eight layers of double sided silicon tracker are arrayed transverse to the magnet axis. The tracker measures the trajectory of relativistic singly charged particles with an overall[a] accuracy of about 17μ in the bending coordinate and 30μ in the non-bending one, as well as provides measurements of the particle energy loss. The time of flight system has four layers, measures singly charged particle transit times with an accuracy of 140 psec and also yields energy loss and coordinate measurements. The TRD is situated on top of the spectrometer and consists of twenty 12 mm thick foam radiator arrays, interleaved by arrays of 6 mm diameter gas proportional tubes filled with Xe/CO_2 mixture. The TRD provides the e^-/hadron separation better than one hundred up to energy 200 GeV as well as precise charged particle coordinate measurements. The RICH detector is installed below the last ToF plane and consists of a 2 cm thick aerogel radiator with refraction index of 1.05, a reflection mirror and pixel type photo-tubes matrix for the light

---

[a]readout pitch size, noise and alignment induced errors summed together.

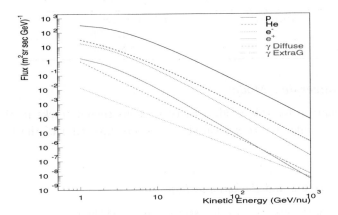

Fig. 2. The assumed cosmic ray fluxes.

detection. It assures the measurement of the velocity of the single charged particle with accuracy better than per mil as well as particle flight direction measurement. The ECAL detector is situated at the bottom of AMS. It is a three-dimensional electromagnetic sampling calorimeter with total length of $15X_0$, consisting of 1 mm diameter scintillating fibers sandwiched between grooved lead plates.

The simulated performance of the AMS subdetectors was checked against test flight and prototype test beam data and found to be in good agreement with them. In addition, 1 to 2 % of the subdetectors readout channels were assumed dead.

The simulated events were triggered by either the coincidence of signals in three out of four ToF planes or large energy deposit in the ECAL[5].

Finally, more than 10^9 events containing p, He, $e^{\pm}$ and γ at different energies have been fully simulated passing through the detector and then reconstructed.

3. Signal and Backgrounds

The dominant cosmic ray components are the p, following by He and C nuclei and then e^-s, representing the backgrounds to the (tiny) e^+ and γ signals. Fig. 2 shows the assumed signal and background fluxes for the AMS energy range.

To get background to signal ratio of the order of a few percent an $O(10^4)$ to $O(10^6)$ background rejection level should be obtained.

4. AMS02 Photon Detection Capabilities

To detect photons in the AMS experiment two complementary methods were explored. The first method or *conversion mode* consisted in the identification and reconstruction of e^+e^- pairs from γ conversions in the material upstream of the first silicon tracker layer, while the second or *single photon mode* was based on the detection of photons in the electromagnetic calorimeter.

Table 1. Rejection factors obtained for various CR particles.

Particles	Rejection Factor
e^-	$> 1. \times 10^5$
p	$> 8. \times 10^5$

4.1. *The Conversion Mode*

The event signature for this mode was two reconstructed tracks in the Tracker coming from a common vertex located somewhere upstream of the first tracker layer[b].

4.1.1. *Event Selection and Background Rejection*

The main issue of the events selection study concerned the pattern recognition and track finding for double track events originating from a common vertex. Optimisation pending, the current results have to be considered preliminary ones. Primary γ-ray energy and incidence direction were determined by adding the fitted momenta vectors of $e^\pm$ pair, evaluated at the entrance of the AMS detector.

The main source of background came from p and e^-, which interacted with the AMS materials, producing secondaries, mainly delta rays within the spectrometer and were reconstructed as double-track events, mimicking the e^+e^- from γ conversion.

The flux of genuine photon events generated in the vicinity of the AMS, namely in the ISS body and solar panels, was found to be small compare to the cosmic one.

The following criteria were applied to reject background events:

- Identify events with interactions;
- Identify charged particles entering the TRD from the top and lighting all the tubes along its reconstructed trajectory.
- Identify particles entering the fiducial volume of the AMS by passing through the side of the TRD (Fig.3.a).
- Identify large reconstructed invariant mass events (Fig.3.b).

Table 1 shows the obtained rejection factors for different cosmic ray species after all cuts[7] have been applied.

4.1.2. *Results*

The AMS-02-γ simulated detection through the pair conversion method, yielding the results as shown in Fig.4. On the first figure the detector acceptance is shown as a function of the γ-ray energy between 1 and 300 GeV. The aperture is a result of three main contributions: the AMS geometrical acceptance, the pair conversion probability, the double-track reconstruction efficiency and event selection efficiency.

---

[b]The material in front of the first silicon tracker plane, consisting of the TRD, the first two layers of ToF scintillators, and mechanical supports, represents $\simeq 0.22 X_0{}^6$.

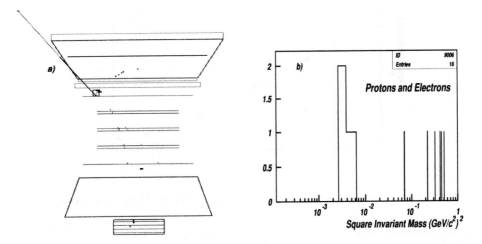

Fig. 3. a) Simulated electron and δ-ray event mimicking a γ converted in a e^+e^--pair at the bottom of the TRD. b) Invariant mass distrubution of few p and e^- survived events.

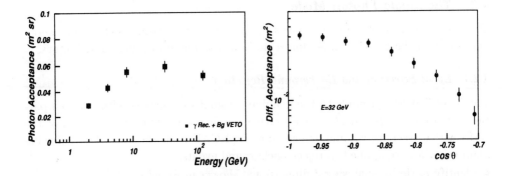

Fig. 4. Conversion mode: AMS acceptance as a function of γ-ray energy (left). Differential acceptance $A(E,\theta)$ versus zenith angle at 32 GeV (right).

Above 200 GeV the $e^\pm$ pair detection was limited by the double-hit resolution of the Tracker. Within 7 to 200 GeV the detector acceptance was at the *plateau* value of 0.058 $m^2 sr$. The AMS-02-γ differential acceptance is shown in Fig.4.(right). It had a maximal opening angle of 45°.

For most of the events energy resolution was dominated by multiple scattering and tracker measurement errors and was parametrised as $\frac{\sigma(E)}{E} = 0.03 \oplus 0.5E(\text{TeV})$. Due to the hard bremsstrahlung radiation by e^+e^- pairs in the AMS materials a sizeable fraction[c] of the photon events had the reconstructed energy differed from the nominal one by more than 5 standard deviations defined above. Finally the

---

[c] 10 to 30 % depending on photon energy

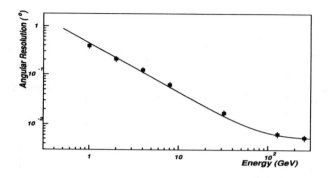

Fig. 5. Conversion mode: AMS angular resolution as a function of the photon energy.

angular resolution (Fig.5) ranged from $0.4°$ to $0.005°$, due mainly to the multiple coulomb scattering. The energy dependence of angular resolution was parametrised as $\sigma(\theta) = 0.005° \oplus \frac{0.44°}{E(\text{GeV})}$.

4.2. The Single Photon Mode

The event signature for this mode was the presence of electromagnetic-type energy deposition in the ECAL, while almost nothing was found in other subdetectors.

4.2.1. Event Selection and Background Rejection

The identified backgrounds to the genuine γ signal were events with charged particles (mostly e^-, p and He nuclei) either passing undetected in the gaps of the AMS active tracking volume or entering the ECAL from the side. The following main criteria were applied to reject background events:
- Identify p, He by analysing 3-dimensional shower in ECAL;
- Identify charged particles by requiring the trajectory direction of the reconstructed ECAL shower to pass inside AMS sensitive volume and reject events with the signals in various AMS subdetectors around it. Fig. 6 shows the distribution of the ToF hits found near the reconstructed particle trajectory for the signal and background events.

Table 2 shows the obtained rejection factors for different cosmic ray species after all cuts[8] have been applied.

Table 2. Rejection factors obtained for various CR particles.

Particles	Rejection Factor
$e^\pm$	$> 53 \times 10^3$
p	$(2.5 \pm 1) \times 10^6$
He Nuclei	$> 1.7 \times 10^6$

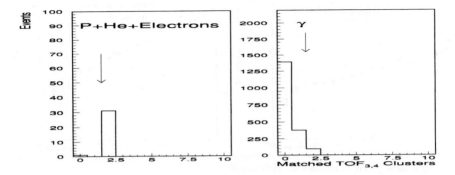

Fig. 6. The Distribution of the number of the ToF hits found near the reconstructed particle trajectory in the two bottom ToF planes for signal (right panel) and background (left panel) events. The arrows show the value of the cut applied.

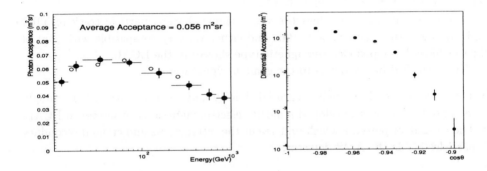

Fig. 7. Left Panel: AMS acceptance as a function of energy. Right Panel: Differential acceptance $A(E,\theta)$ versus zenith angle at 50 GeV.

4.2.2. *Results*

Fig. 7 shows the obtained average and differential acceptances after all cuts have been applied. The average acceptance is situated around 0.58 m^{2}sr with maximal open angle around 22°. The relative energy resolution was parametrised as $0.03 \oplus \frac{0.13}{\sqrt{E(GeV)}}$ up to 2 TeV, and angular resolution as $0.6° \oplus \frac{4.5°}{\sqrt{E(GeV)}}$.

4.3. *A Combined Example*

Combining the above two photon detection methods, we simulated the AMS 3 year measurement of the photon spectra for Galactic and ExtraGalactic diffuse spectra[9], which is shown in Fig. 8.

5. AMS02 Positron Detection Capabilities

To detect e^+ in the AMS experiment two methods were explored. The common part of the two methods was to use TRD for e/h discrimination and ToF/Tracker

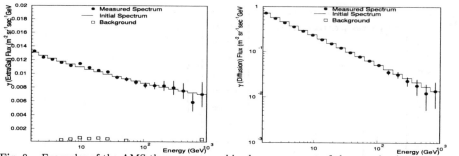

Fig. 8. Examples of the AMS three years combined measurement of the cosmic photon spectra: Galactic (right panel) and ExtraGalactic (left panel).

for charge and sign of charge determination. The first method or *low energy mode* then relied on RICH velocity measurement while the second one or *high energy mode* used ECAL to further identify e^+.

The e^+ event signature therefore was a single charged particle track in TRD and Tracker with γ factor above 1000 and either velocity compatible with that of light or have developed electromagnetic-type shower in the ECAL.

The identified backgrounds to the genuine signal were:

- The e^- events with wrongly measured rigidity and p events with either wrongly measured velocity or developed a mostly electromagnetic-type shower in ECAL.
- The e^- and/or p events which underwent the interactions and created secondary particles inside the AMS detector.

5.1. *Event Selection and Background Rejection*

To partly remove the p background the TRD was used. Fig. 9 shows TRD signal distribution for e^- and p. The obtained TRD e/h rejection ranged from 10^3 for 10

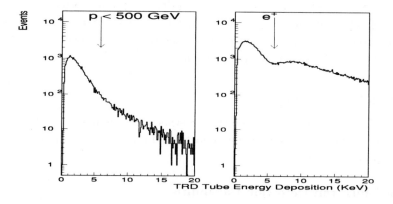

Fig. 9. The distribution of the TRD signals for p (left panel) and e^-s (right panel). Arrows show the value of the cut applied.

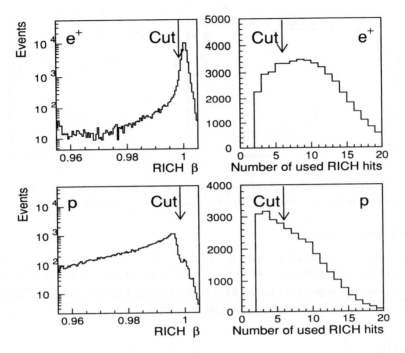

Fig. 10. The distribution of the particle velocity measured by RICH and the number of photon hits used in the measurement, for signal (upper panels) and background (lower panels) events. The arrows show the value of the cut applied.

GeV p to 10^2 for 300 GeV p. In addition, vetoing events with additional signals in the AMS subsystem in the vicinity of the reconstructed particle trajectory effectively removed the the bulk of the e^- background.

5.1.1. *Low Energy Mode*

The AMS RICH velocity measurement can be used to identify $e^\pm$ with momentum less than 12.5 GeV/c. To provide high discrimination power against p background the method was applied to particles with momentum below 11 GeV/c.

The following are the main selection criteria required to reject background events:

- Identify e^+ as particles with track charge +1 and the velocity measured by RICH compatible to 1. Fig. 10 shows the distributions of the measured velocity and number of photons used to reconstruct the Cherenkov ring for signal and background events and the values of applied cuts ;
- Require the particle trajectory to pass inside the AMS sensitive volume and reject events with the signals in various AMS subdetectors around it;

Fig. 11 shows the estimated background rejection factor after all cuts[10].

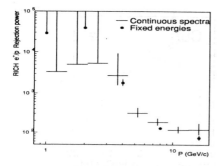 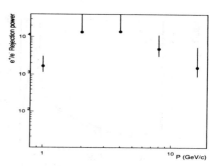

Fig. 11. Low energy mode: The obtained rejection power for the e^+ signal 1-11 GeV/c as a function of particle momentum against p (left panel) and against e^- (right panel).

5.1.2. *High Energy Mode*

To further reduce the p background the 3-dim analysis of the ECAL energy distribution was done. Finally, the residual e^- and/or p background was removed by matching the energy deposition in the ECAL and the rigidity measured by Tracker. Fig. 12 shows the obtained rejection against e^- and p.

5.2. *Results*

5.2.1. *Low Energy Mode*

The estimated average and differential acceptance is shown on Fig. 13. The relative energy resolution turned out to be around 2%, the 20-25% low energy tail being observed due to hard photon radiation in the AMS materials.

5.2.2. *High Energy Mode*

Fig. 14 shows the estimated e^+ average and differential acceptance after all cuts[11] had been applied. The e^+ energy estimation was done by using combined ECAL

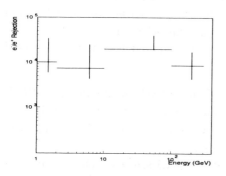 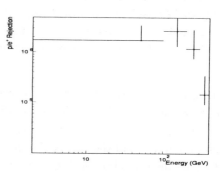

Fig. 12. High energy mode: The obtained e^+ rejection as a function of particle momentum against e^- (left panel) and p (right panel).

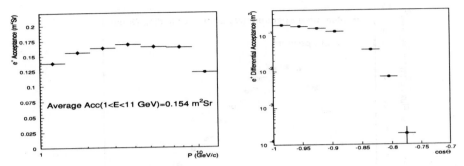

Fig. 13. Estimated acceptance for the e^+ as a function of particle momentum (left panel) and differential acceptance at 1 GeV/c (right panel).

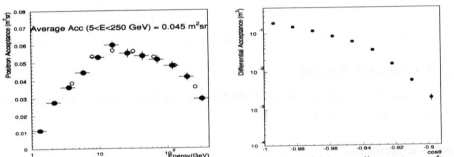

Fig. 14. The estimated e^+ average (left panel) and differential (right panel) acceptance after all cuts had been applied.

energy and Tracker rigidity quantities, the resolution turning out to be similar to the photon one for high energies with some improvement for the low ones. Fig. 15 shows the reconstructed energy distribution for e^+ of 16 GeV.

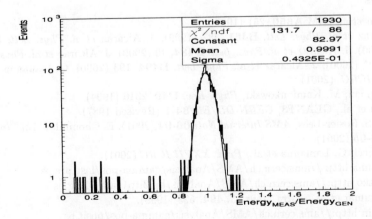

Fig. 15. The reconstructed relative energy distribution for 16 GeV e^+.

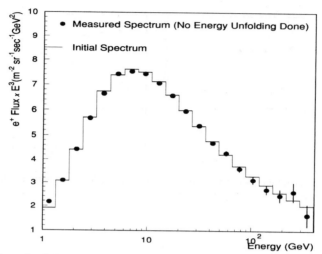

Fig. 16. Example of the AMS three years combined measurement of the cosmic e^+ spectrum.

5.3. *A Combined Example*

Fig. 16 shows the simulation of the AMS three years high statistics measurement of cosmic e^+ spectrum up to 400 GeV.

Acknowledgements

We are grateful to our AMS colleagues for the numerous fruitful discussions concerning the AMS02 subdetector system performance. We thank Dr. A. Biland and Dr. C. Goy for providing us most of the MC event samples used in this work. We thank Dr. E. Choumilov for the invaluable help in understanding the simulation details of the ToF and ECAL detectors.

References

1. S. Ahlen *et al.* NIM **A350**, 251 (1994).
2. J. Alcaraz *et al. Phys. Lett.* **B461**, 387 (1999). J. Alcaraz *et al. Phys. Lett.* **B472**, 215 (2000). J. Alcaraz *et al. Phys. Lett.* **B484**, 10 (2000). J. Alcaraz *et al. Phys. Lett.* **B490**, 27 (2000). J. Alcaraz *et al. Phys. Lett.* **B494**, 193 (2000). G. Lamanna, *Proc. XXVII ICRC* (2001).
3. G. Jungman, M. Kamionkowski, *Phys. Rev.* **D49**, 2316 (1994).
4. R. Brun et al., GEANT3, *CERN-DD/EE/84-1* (Revised 1987).
5. C. Goy, S. Rosier-Lees, *AMS Internal Note* **06-04**,(2001). E. Choumilov *AMS Internal Note* **06-05**,(2001).
6. B. Bertucci, G. Lamanna et al., *Proc. XXVII ICRC* (2001).
7. G.Lamanna http://ams.cern.ch/AMS/Analysis/gamma-pos/last_pre.ps
8. V.Choutko http://ams.cern.ch/AMS/Analysis/hpl3itp1/ams02_sg.ps
9. P. Sreekumar et al., *Astro. Jour. S.* **494**, 523 (1998).
10. A.Malinin http://ams.cern.ch/AMS/Analysis/gamma-pos/posit.ps
11. V.Choutko http://ams.cern.ch/AMS/Analysis/hpl3itp1/ams02_pos_ecal.ps

DARK MATTER SEARCH WITH GAMMA RAYS: THE EXPERIMENTS EGRET AND GLAST

ALDO MORSELLI*

Dept. of Physics, Univ. of Roma "Tor Vergata" and INFN Roma 2, Roma, Italy

The direct detection of annihilation products in cosmic rays offers an alternative way to search for supersymmetric dark matter particles candidates. The study of the spectrum of gamma-rays, antiprotons and positrons offers good possibilities to perform this search in a significant portion of the Minimal Supersymmetric Standard Model parameters space. In particular the EGRET team have seen a convincing signal for a strong excess of emission from the galactic center that have not easily explanation with standard processes. We will review the achievable limits with the experiment GLAST taking into accounts the LEP results and we will compare this method with th antiproton and positrons experiments, the direct underground detection and with future experiments at LHC.

Keywords: gamma-rays; dark matter; supersymmetry

1. the EGRET data

The EGRET team [1] have seen a convincing signal for a strong excess of emission from the galactic center, with I(E) x E^2 peaking at 2 GeV, and in an error circle of 0.2 degree radius including the position l = 0° and b = 0°. In figure 1 is shown the map towards the galactic center.

This is a particular aspect of a more general problem of the diffuse Galactic gamma-ray emission [2] that also outside the galactic center reveal a spectrum which is harder than expected. As can be seen in figure 2, the spectrum observed with EGRET below 1 GeV is in accord with the assumption that the cosmic ray spectra and the electron-to-proton ratio observed locally are uniform, however, the spectrum above 1 GeV, where the emission is supposedly dominated by $\pi°$-decay, is harder than that derived from the local cosmic ray proton spectrum [3]. Many differen approach are trying to solve the problem, as the realiving of the assumption that the local cosmic ray electron spectra is not representative for the Galaxy and it is in average harder than that measured locally, or dispersion in the cosmic ray source spectra such that the SNR would produce power-law spectra with varying indices (for a discussion see [4]). Here we will connect the problem of the GeV excess with the problem of the missing dark matter in the Universe and we will examine

---

*E-mail: aldo.morselli@roma2.infn.it

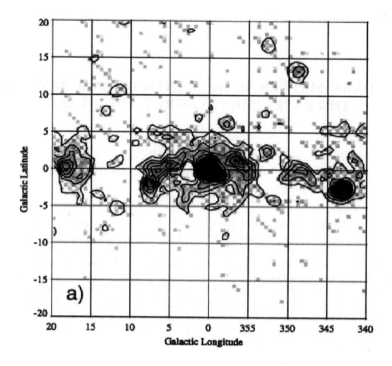

Fig. 1. *Residual smoothed profiles owards the galactic center after subtraction of the model-predicted diffuse emission background for $E > 1$ GeV*

the possibility to disantangle this effect with the future space γ-ray and cosmic ray experiments.

Over the last years our knowledge of the inventory of matter and energy in the universe has improved dramatically. Astrophysical measurements from disparate experiments are now converging and a standard cosmological model is emerging. The most significant new data come from recent measurements of the cosmic microwave background radiation (CMBR)[5] and measurements of the Hubble flow using distant supernovae[6].

The evidence currently favors (see for example [7]) a flat universe with a cosmological constant $\Omega_\Lambda = 1 - \Omega_m$ and a total matter density of about 40%±10% of the critical density of the Universe, with a contribution of the baryonic dark matter less then 5% and a contribution from neutrinos that cannot be greater then 10%. The remaining matter should be composed of yet-undiscovered Weakly Interacting Massive Particles (WIMP), and a good candidate for WIMP's is the Lightest Supersymmetric Particle (LSP) in R-parity conserving supersymetric models.

The motivation for supersymmetry at an accessible energy is provided by the gauge hierarchy problem [8], namely that of understanding why $m_W \ll m_P$, the only candidate for a fundamental mass scale in physics. This difference introduce problems because one must fine-tune the bare mass parameter so that is almost

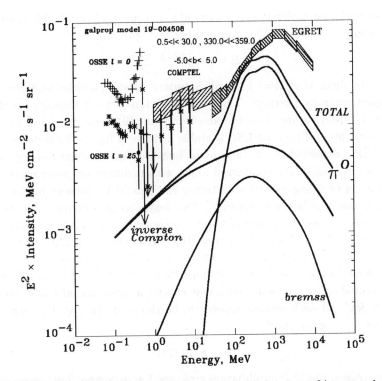

Fig. 2. *Gamma-ray energy spectrum of the inner galaxy ($300° \geq l \leq 30°$) compared with what is expected for standard propagation models* [3].

exactly cancelled by the quantum correction in order to obtain a small physical value of m_W. This seems unnatural, and the alternative is to introduce new physics at the TeV scale and to postulate approximate supersymmetry[9], whose pairs of boson and fermions produce naturally cancelling quantum corrections that are naturally small if

$$|m_B^2 - m_F^2| \leq 1 TeV$$

This is also the reason to expect that, if supersymmetry is real, it might be accessible to the current generation of accelerators and in the range expected for a cold dark matter particle.

The minimal supersymmetric extension of the Standard Model (MSSM) [10] has the same gauge interactions as the Standard Model and has the advantage that all the phenomenology can be parametrized by five parameters: the higgs mixing parameters μ that appears in the neutralino and chargino mass matrices, the common mass for scalar fermions at the GUT scale m_0, the gaugino mass parameter $M_{1/2}$, the trilinear scalar coupling parameter A and the ratio between the two vacuum expectation values of the Higgs fields defined as $\tan \beta = v_2/v_1 = <H_2>/<H_1>$.

The LSP is expected to be stable in the MSSM, and hence should be present in the Universe today as a cosmological relic from the Big Bang [11]. This is a consequence of a multiplicatively-conserved quantum number called R-parity, which is related to baryon number, lepton number and spin:

$$R = (-1)^{3B+L+2S}$$

It is easy to check that R=+1 for all Standard Model particles and R=-1 for all their supersymmetric partners. There are three important consequences of R conservation: (i) sparticle are always produces in pairs; (ii) heavier sparticles decay into lighter sparticles and (iii) the LSP is stable because it has no legal decay mode.

The LSP is expected also to be neutral, because with an electric charge or strong interaction, it would have condensed along with ordinary baryonic matter during the formation of astrophysical structures, and should be present in the Universe today in anomalous heavy isotopes [12]. This leaves as candidates a sneutrino with spin 0 , the gravitino with spin 2/3 and the neutralino χ that is a combination of the partners of the γ, Z and the neutral Higgs particles (spin 1/2).

The sneutrino seems to be ruled out by searches for the interactions of relic particles with nuclei that require a sneutralino mass greater then few TeV [13] while the gravitino could constitute warm dark matter with a mass around 1 keV. So the best candidate for cold dark matter appears to be the neutralino χ. The experimental LEP lower limit on m_χ is [14]

$$m_\chi \geq 50 \; GeV$$

As m_χ increases, the LSP annihilation cross section decreases, but, as we will show below, up to $\sim 400 \; GeV$ a possible signature of the existence of the LSP is a bump in the spectrum of the diffuse gamma ray background around the neutralino mass due to neutralino annihilation in the halo [15]. The bump arise because if neutralinos make up the dark matter of our galaxy, they would have non-relativistic velocities.

How can be see this kind of signal ? In the next session we present one possibility, i.e. the experiment GLAST.

2. The Gamma-ray Large Area Telescope GLAST

The standard techniques for the detection of gamma-rays in the pair production regime energy range are very different from the X-ray detection. For X-rays detection focusing is possible and this permits large effective area, excellent energy resolution, very low background. For gamma-rays no focusing is possible and this means limited effective area, moderate energy resolution, high background but a wide field of view. This possibility to have a wide field of view is enhanced now, in respect to EGRET, with the use of silicon detectors, that allow a further increase of the ratio between height and width, essentially for two reasons: a) an increase of the position resolution that allow a decrease of the distance between the planes of the tracker without affect the angular resolution, b) the possibility to use the silicon detectors themselves for the trigger of an events, with the elimination of the Time of Flight system, that require some height.

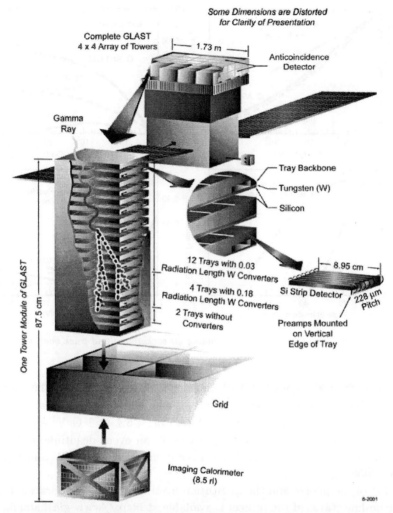

Fig. 3. The *GLAST* instrument, exploded to show the detector layers in a tower, the stacking of the CsI logs in the calorimeter, and the integration of the subsystems.

The Gamma-ray Large Area Space Telescope (GLAST)[16], has been selected by NASA as a mission involving an international collaboration of particle physics and astrophysics communities from the United States, Italy, Japan, France and Germany for a launch in the first half of 2006. The main scientific objects are the study of all gamma ray sources such as blazars, gamma-ray bursts, supernova remnants, pulsars, diffuse radiation, and unidentified high-energy sources. Many years of refinement has led to the configuration of the apparatus shown in figure 3, where one can see the 4x4 array of identical towers each formed by: • Si-strip Tracker Detectors and converters arranged in 18 XY tracking planes for the measurement of the photon direction. • Segmented array of CsI(Tl) crystals for the measurement

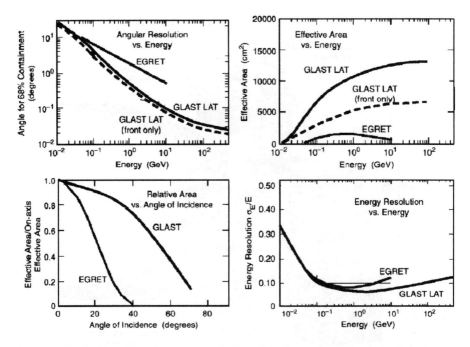

Fig. 4. *Instrument performance, including all background and track quality cuts.*

the photon energy. • Segmented Anticoincidence Detector (ACD). The main characteristics, shown in figure 4, are an energy range between 20 MeV and 300 GeV, a field of view of ~ 3 sr, an energy resolution of $\sim 5\%$ at 1 GeV, a point source sensitivity of 2×10^{-9} (ph cm^{-2} s^{-1}) at 0.1 GeV, an event deadtime of 20 μs and a peak effective area of 10000 cm^2, for a required power of 600 W and a payload weight of 3000 Kg.

The list of the people and the Institution involved in the collaboration together with the on-line status of the project is available at *http://www-glast.stanford.edu*. A description of the apparatus can be found in [17] and a description of the main physic items can be found in [18].

GLAST is particularly interesting for the supersymmetric particle search because, if neutralinos make up the dark matter of our galaxy, they would have non-relativistic velocities, hence the neutralino annihilation into the gamma gamma and gamma Z final states can give rise to gamma rays with unique energies $E_\gamma = M_\chi$ and $E'_\gamma = M_\chi (1 - m_z^2/4M_\chi^2)$.

In figure 5 is shown how strong can be the signal[19] in the case of a cuspy dark matter halo profiles distribution[20]. Figure 6 shows the GLAST capability to probe the supersymmetric dark matter hypothesis[19]. The various zone sample the MSSM with different values of the parameters space for three classes of neutralinos. The previous galaxy dark matter halo profile[20] that gives the maximal flux has been assumed. The solid line shows the number of events needed to obtain a 5 σ detection

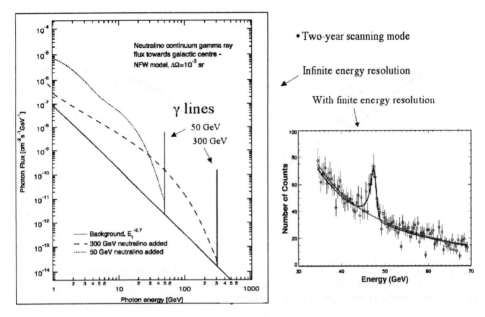

Fig. 5. *Total photon spectrum from the galactic center from $\chi\chi$ annihilation (on the left), and number of photons expected in GLAST for $\chi\chi \to \gamma\gamma$ from a 1-sr cone near the galactic center with a 1.5 % energy resolution (on the right)*

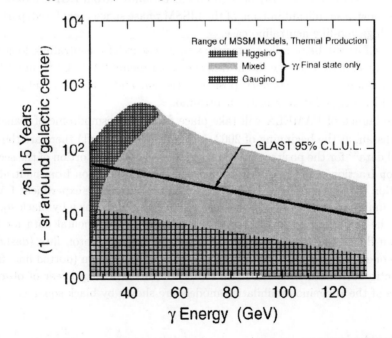

Fig. 6. *Number of photons expected in GLAST for $\chi\chi \to \gamma\gamma$ from a 1-sr cone near the galactic center as a function of the possible neutralino mass. The solid line shows the number of events needed to obtain a five sigma signal detection over the galactic diffuse gamma-ray background as estimated by EGRET data.*

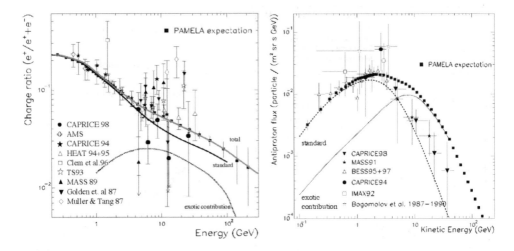

Fig. 7. *Distortion of the secondary positron fraction (on the left) and secondary antiproton flux (on the right) induced by a signal from a heavy neutralino. The PAMELA expectation in the case of exotic contributions are shown by black squares*

over the galactic diffuse γ-ray background as estimated from EGRET data. As the figures show, a significant portion of the MSSM phase space is explored, particularly for the higgsino-like neutralino case.

This effort will be complementary to a similar search for neutralinos looking with cosmic-ray experiments like the next space experiments PAMELA[21] and AMS[22] at the distortion of the secondary positron fraction and secondary antiproton flux induced by a signal from a heavy neutralino.

The launch of PAMELA will take place from the cosmodrome of Baikonur, in Kazakhstan, at the beginning of 2003. In figure 7 (on the left) there are the experimental data[23] for the positron fraction together with the distortion of the secondary positron fraction (solid line) due to one possible contribution from neutralino annihilation (dotted line, from[24]). The expected data from the experiment PAMELA in the annihilation scenario for one year of operation are shown by black squares[25].

In the same figure (on the right) there are the experimental data for the antiproton flux[26] together with the distortion on the antiproton flux (dashed line) due to one possible contribution from neutralino annihilation (dotted line, from[27]). The antiproton data that PAMELA would obtain in a single year of observation for one of the Higgsino annihilation models are shown by black squares.

3. Conclusion

The gamma-ray space experiment GLAST is under construction. Its time of operation and energy range is shown together with the other space X-ray satellite

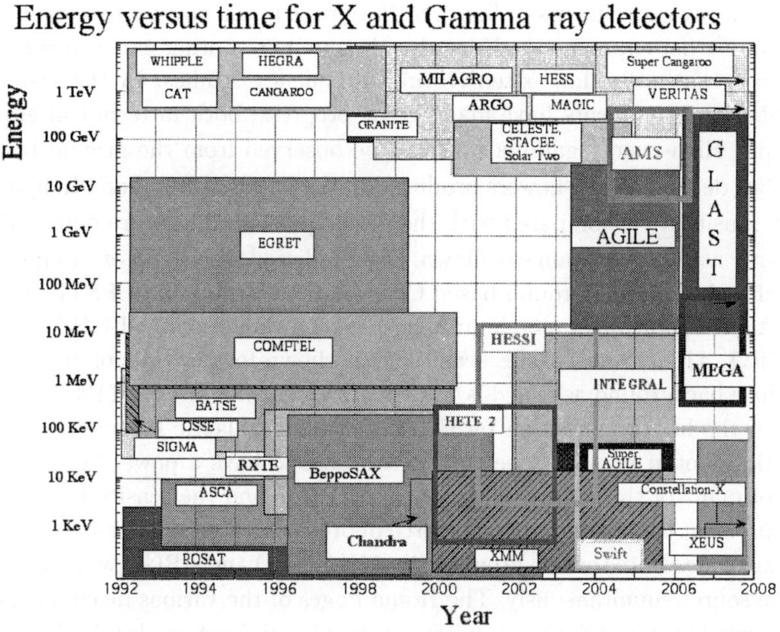

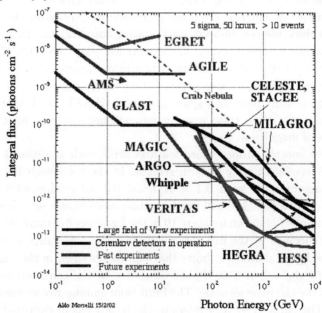

Fig. 8. *Timeline schedule versus the energy range covered by present and future detectors in X and gamma-ray astrophysics.*

Fig. 9. *Sensitivity of present and future detectors in the gamma-ray astrophysics.*

and gamma-ray experiments in figure 8. Note that it will cover an interval not covered by any other experiments. Note also the number of other experiments in other frequencies that will allow extensive multifrequency studies. In the last decade, ground-based instruments have made great progress, both in technical and scientific terms. High-energy gamma rays can be observed from the ground by experiments that detect the air showers produced in the upper atmosphere. In figure 9 the GLAST sensitivity is compared with the others present and future detectors in the gamma-ray astrophysics range is shown. The predicted sensitivity of a number of operational and proposed Ground based Cherenkov telescopes, CELESTE, STACEE, VERITAS, Whipple is for a 50 hour exposure on a single source. EGRET, GLAST, MILAGRO, ARGO and AGILE sensitivity is shown for one year of all sky survey. The diffuse background assumed is $2 \cdot 10^{-5}$ $photons$ $cm^{-2}s^{-1}sr^{-1}(100\ MeV/E)^{1.1}$, typical of the background seen by EGRET at high galactic latitudes. The source differential photon number spectrum is assumed to have a power law index of -2, typical of many of the sources observed by EGRET and the sensitivity is based on the requirement that the number of source photons detected is at least 5 sigma above the background. Note that on ground only MILAGRO and ARGO will observe more than one source simultaneously. The Home Pages of the various instruments are at http://www-hfm.mpi-hd.mpg.de/CosmicRay/CosmicRaySites.html. The arrow for AMS indicates that a published estimate does not exist but flux sensitivity should be of the order of $2 \cdot 10^{-8}$ $photons$ $cm^{-2}s^{-1}$ above few Gev [28].

A wide variety of experiments provide interesting probes for the search of supersymmetric dark matter. Indirect dark matter searches and traditional particle searches are highly complementary. In the next five years, an array of experiments will be sensitive to the various potential neutralino annihilation products. These include under-ice and underwater neutrino telescopes, atmospheric Cerenkov telescopes and the already described space detectors GLAST and PAMELA together with AMS. In many cases, these experiments will improve current sensitivities by several orders of magnitude.

Direct dark matter probes share features with both traditional and indirect searches, and have sensitivity in both regions. In the cosmologically preferred regions of parameter space with $0.1 < \Omega_\chi h^2 < 0.3$, all models with charginos or sleptons lighter than 300 GeV will produce observable signals in at least one experiment. An example[29] is shown in figure 10 in the framework of minimal supergravity, which is fully specified by the five parameters $m_0, M_{1/2}, A_0, \tan\beta, \text{sgn}(\mu)$ defined in section 1. The figure shows the limits that can be obtained in the $m_0, M_{1/2}$ plane for $\tan\beta = 10$, $A_0 = 0$, $\mu > 0$. Higher values (~ 50) of $\tan\beta$ requires significant fine-tuning of the electroweak scale. The limit from gamma-ray assumes a moderate halo profile. The curve $B \to X_s \gamma$ refers to the improvement expected for the same date from BaBar, BELLE and B factories in respect to the CLEO and ALEPH results[30]. The curve $\Phi_\mu^\odot$ refers to the indirect DM search with underwater ν experiments like AMANDA, NESTOR and ANTARES[31] and the curve σ_p refers to the direct DM search with underground experiments like DAMA, CDMS, CRESST

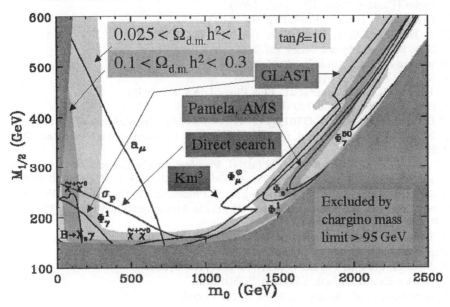

Fig. 10. *Example of estimated reaches of various searches before the LHC begins operation. Note the complementarity between the different techniques. For moderate values of* $\tan \beta$ *all the cosmological interesting region will be covered (see text for details).*

and GENIUS[32]

Acknowledgements

I would like to thanks all the component of the GLAST Dark Matter working group for lots of discussion on the argument, in particular Piero Ullio, Elliot Bloom, Eduardo do Couto e Silva and Andrea Lionetto; Hans Mayer-Hasselwander for discussion on the EGRET data from the galactic center; Roberto Battiston for the organization of this very nice workshop and discussions.

References

1. H. Mayer-Hasselwander *et al.*, Astron. Astrophys. 335, 161 (1998).
2. S. Hunter *et al.*, *Astrophys. J.* **481**, 205(1997).
3. A. Strong *et al.*, *Astrophys. J.* **537**, 763(2000).
4. M. Pohl, astro-ph/0111552, (2001).
5. P. de Bernardis *et al.*, Frascati Physics Series Vol.XXIV, 399,(2002), http://www.roma2.infn.it/infn/aldo/ISSS01.html.
6. A. Riess *et al.*, *Astrophys. J.* **560**, 49(2001) [astro-ph/0104455].
7. J. Primack, Frascati Physics Series Vol.XXIV, 449, (2002), http://www.roma2.infn.it/infn/aldo/ISSS01.html [astro-ph/0112255].

8. L. Maiani, *Proc. Summer School on Particle Physics*, Gif-sur-Yvette, 1979 (IN2P3, Paris, 1980), 3. G. 't Hooft in G. 't Hooft et al., eds., *Recent Developments in Field Theories* (Plenum Press, New York, 1980).
9. P. Fayet and S. Ferrara, Phys. Rep. 32, 251 (1977).
10. H. E. Haber and G. L. Kane, Phys. Rep. 117, 75 (1985).
11. J. Ellis, Frascati Physics Series Vol. XXIV, 49 (2002), http://www.roma2.infn.it/infn/aldo/ISSS01.html.
12. P. Smith, Contemp. Phys. 29, 159 (1998).
13. H. Klapdor-Kleingrothaus et al., Eur. Phys. J. **A3**, 85 (1998).
14. J. Ellis et al., hep-ph 0004169 (2000).
15. V. Berezinsky, Phys. Lett. **B261**, 71 (1991); A. Morselli, *The dark side of the Universe*, p. 267 (1994), World Sci. Co. G. Jungman and M. Kamionkowski, Phys. Rev. **D51**, 3121 (1995).
16. W. Atwood et al., NIM **A342**, 202 (1994). Proposal for the Gamma-ray Large Area Space Telescope, SLAC-R-522 (1998). B. Dingus et al., 25th ICRC, OG 10.2.17, **5**, p. 69, Durban. A. Morselli, Very High Energy Phenomena in the Universe, Ed. Frontiers, 123 (1994). A. Morselli, "Frontier Objects in Astrophysics and Particle Physics," SIF, Bologna **65**, 613 (1999).
17. R. Bellazzini, Frascati Physics Series, Vol. XXIV, 353 (2002), http://www.roma2.infn.it/infn/aldo/ISSS01.html.
18. A. Morselli, Frascati Physics Series, Vol. XXIV, 363 (2002), http://www.roma2.infn.it/infn/aldo/ISSS01.html.
19. L. Bergstrom et al., Astropart. Phys. **9**, 137 (1998).
20. J. Navarro et al., Astrophys. J. **462**, 563 (1994).
21. P. Spillantini et al., 24th ICRC Roma, OG 10.3.7, 591 (1995). V. Bonvicini et al., NIM **A461**, 262 (2001). P. Spillantini, Frascati Physics Series, Vol. XXIV, 249 (2002), http://www.roma2.infn.it/infn/aldo/ISSS01.html.
22. R. Battiston, Frascati Physics Series, Vol. XXIV, 261 (2002), http://www.roma2.infn.it/infn/aldo/ISSS01.html and references therein.
23. M. Boezio et al., Astrophys. J. **532**, 653 (2000) and references therein.
24. E. Baltz and J. Edsjö, Phys. Rev. **D59**, 023511 (1999).
25. P. Picozza and A. Morselli, The Ninth Marcel Grossmann Meeting, World Scientific (2001) [astro-ph/0103117] and references therein.
26. D. Bergström et al., ApJ Letters **534**, L177 (2000) and references therein.
27. P. Ullio, astro-ph/9904086 (1999).
28. R. Battiston et al., Frascati Physics Series, Vol. XXIV, 381 (2002), http://www.roma2.infn.it/infn/aldo/ISSS01.html and references therein.
29. L. Feng et al., Phys. Rev. **D63**, 045024 (2001) [astro-ph/0008115].
30. A. L. Kagan and M. Neubert, Eur. Phys. J. **C7**, 5 (1999) [hep-ph/9805303].
31. AMANDA Collaboration, E. Dalberg et al., HE.5.3.06, 26th ICRC (1999). C. Spiering et al., astro-ph/9906205. NESTOR Collaboration, http://www.uoa.gr/nestor. ANTARES Collaboration, J. R. Hubbard et al., HE.6.3.03, 26th ICRC (1999).
32. DAMA Collaboration, R. Bernabei et al., Phys. Lett. **B480**, 23 (2000). CDMS Collaboration, R. W. Schnee et al., Phys. Rept. **307**, 283 (1998). CRESST Collaboration, M. Bravin et al., Astropart. Phys. **12**, 107 (1999) [hep-ex/9904005].